PREFACE

To the Student

This study guide is designed to accompany *Understanding the Human Body* by Tate, Seeley, and Stephens. Each chapter in the study guide, and the order of topics within the chapter, corresponds to a chapter in the text. This makes it possible for you to study systematically and also makes it easier for you to find or to review information. Read the chapter material in the text before you use this study guide. It is designed to help you understand and master the subject of anatomy and physiology.

FEATURES

Focus

Each chapter begins with a focus statement, briefly reviewing some of the main points of the text chapter. This is not a chapter summary; you will find that in the text. The chapter summary is useful. Reading it should become a routine part of your study habits. In the study guide, the focus statement sets the stage by reminding you of the major concepts you should have learned by reading the textbook.

Content Learning Activity

This section of the study guide contains a variety of exercises including matching, completion, ordering, and labeling activities, arranged by order of topics in the text. Each part begins with a quotation from the text or a statement that identifies the subject to be covered. Occasionally, you will find a "bulletin" statement describing important information. Just because that information is not a question does not mean it is unimportant. Quite the contrary. The "bulletin" statements are added to the study guide because they will help you to understand the material, so pay attention to them.

The content learning activity is not a test; it is a strategy to help you learn. Don't guess! If you learn something incorrectly it is difficult to relearn it correctly. Use the textbook or your lecture notes for help whenever you are not sure of an answer. The emphasis here is on learning the content, hence the name of this section. The content questions cover the material in the same sequence as it is presented in the text. Learning the material in this order makes it easier to relate pieces of information to each other, and makes it easier to remember the information.

After completing the exercises check your answers against the answer key. If you missed the correct answer to a question, check the text to make sure you now understand the correct answer. Before going onto the next section of the study guide, review this section to be sure you understand and remember the content. Cover the answers you have written with a piece of paper and mentally answer each exercise once more as you review.

Quick Recall

The quick recall section asks you to list, name, or briefly describe some aspect of the chapter's content. Although this section can be completed rapidly, do not confine yourself to quickly writing down the answers. As you complete each quick recall question, use it to trigger more information in your mind. For example, if the quick recall question asks you to name the major regions of the body, do that, then think of their definition, what their various sub-parts are, visualize them, and so on. This section should be enjoyable and satisfying because it will demonstrate that you have learned the basic information about the material. Verify your answers against the answer key.

Mastery Learning Activity

The mastery learning activity lets you see what you have learned and if you can use that information. It consists of multiple choice questions that are similar to the questions on the exams you will take for a grade, so it is really a "practice" test and should be taken as a test. However, don't guess. This "practice" test is also a learning tool. If you don't know the answer for sure, admit it and then find out what the correct answer is. Some of the questions require recall of information. Others may state the information somewhat differently than the way it appeared in either the text or study guide. This is entirely fair, because in real life you must be able to recognize the information no matter how it is reworded, and you should even be able to express the information in your own words. Another goal of this section is to make you think about the relationship between different bits of information or concepts, so some of the questions are more complex than those requiring only recall. Finally, some questions in this section ask you to use what you have learned to solve new problems.

After you have answered these questions, check the answer key. In addition to the answers, there is a detailed explanation of why a particular answer was correct. Sometimes an explanation of why a choice is incorrect is also given. These explanations are provided because this section is more difficult than the preceding sections. Make sure you understand why each answer is correct. Check the textbook, ask another student, talk with your instructor, but make sure you know. The mastery learning activity will show you the areas that you need to concentrate on further. Use it to improve your understanding of anatomy and physiology.

The format of this section allows you to write the answers to the questions beside each question. If you cover the answers, you can retake the test. Don't be satisfied until you get at least 90% of the questions correct.

Final Challenges

This section of the study guide corresponds to the concept questions at the end of each chapter in the text. These questions challenge you to apply information to new situations, analyze data and come to conclusions, synthesize solutions, and evaluate problems. Some of the problem-solving questions in the mastery learning section have given you practice for the questions in this section. In addition, explanations are provided to help you see how to go about solving questions of this type. Even though explanations are given, write down your answers to the questions on a separate piece of paper. Writing is a good way to organize your thoughts and most of us can benefit from practice in writing. A good way to see if you have communicated your thoughts effectively is to have another student read your answers and see if they make sense.

The questions contain useful information, but they are not designed primarily to help you learn specific information. Rather, they emphasize the thought processes necessary to solve problems. If all you do is read the question and quickly look up the answer, you have defeated the purpose of this section. Think about the questions and develop your reasoning skills. Long after you have forgotten a particular bit of information, these skills will be useful, not only for anatomy and physiology related problems, but for many other aspects of your life as well. We hope that you not only see the benefit of possessing problem-solving skills, but will come to appreciate that solving problems is fun!

A Final Thought

Good luck with all aspects of the anatomy and physiology course you are about to begin. For most students this is a challenging course and we hope that the study guide makes things a little easier and a little clearer. We are confident that when you have completed the course you will be proud of what you have accomplished. Just remember to enjoy the learning process as you go along.

Philip Tate
James Kennedy

ACKNOWLEDGEMENTS

The development and production of this study guide involved much more than the work of the authors and we gratefully acknowledge the assistance of the many other individuals involved. We wish to thank our families for their support, for their encouragement and understanding from the beginning to the end of the project helped to sustain our efforts. Erica Michaels contributed to the development of the design for the study guide. We also wish to acknowledge Laura Edwards and Bob Callanan at Mosby-Year Book College Publishing for their assistance in making the study guide a unique and valuable asset for the student. Our thanks to everybody involved with the artwork taken from the textbook, and to D. Michael Dick for his original study guide illustrations. We also wish to recognize the contribution of Sue Pepe to the design and layout of the study guide. To the reviewers listed below, our gratitude for their thorough and thoughtful critiques, which resulted in a significant improvement of the study guide. Thank you.

STUDY GUIDE to accompany

Understanding the

HUMAN BODY

Phil Tate, D.A.
(Biological Education)

Instructor of Anatomy and Physiology
Phoenix College
Maricopa Community College District
Phoenix, Arizona

James Kennedy, D.A.
(Biological Education)

Instructor of Anatomy and Physiology
Phoenix College
Maricopa Community College District
Phoenix, Arizona

 Mosby

St. Louis   Baltimore   Boston   Chicago   London   Madrid   Philadelphia   Sydney   Toronto

Mosby

Dedicated to Publishing Excellence

Editor-in-chief: James M. Smith
Editor: Robert J. Callanan
Developmental Editor: Laura J. Edwards
Project Manager: Gayle Morris
Production Editor: Lisa R. Nomura
Senior Book Designer: Susan Lane

Printed in the United States of America
Printing/binding by Plus Communications

Mosby-Year Book, Inc.
11830 Westline Industrial Drive
St. Louis, Missouri, 63146-3318

International Standard Book Code Number 0-8016-7198-1

94 95 96 97 98 / 9 8 7 6 5 4 3 2 1

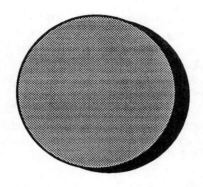

Contents

Introduction to The Human Body

FOCUS: The human organism is often examined at seven structural levels: chemical, organelle, cell, tissue, organ, organ system, and the organism. Anatomy examines the structure of the human organism, and physiology investigates its processes. Structures and processes interact to maintain homeostasis primarily through negative-feedback mechanisms.

CONTENT LEARNING ACTIVITY

Anatomy and Physiology

❝*Knowledge of anatomy and physiology is beneficial to health professionals and nonprofessionals.***❞**

Match these terms with the correct statement or definition:

Anatomy
Physiology

_____ 1. Scientific study of the body's structure.

_____ 2. Considers the relationship between the structure of a body part and its function.

_____ 3. Scientific study of the processes or functions of living things.

_____ 4. Predicts the body's responses to stimuli and considers how the body maintains conditions within a narrow range of values in the presence of a continually changing environment.

Structural and Functional Organization

"*The body can be studied at seven structural levels.***"**

A. **M**atch these terms with the correct statement or definition:

Cell	Organelle
Chemical	Organ system
Organ	Tissue
Organism	

_____ 1. Structure within a cell that performs one or more specific functions.

_____ 2. Basic living units of all plants and animals.

_____ 3. Group of cells with similar structure and function plus the extracellular substances located between them.

_____ 4. Group of organs classified as a unit because of a common function or set of functions.

B. **M**atch these terms with the correct statement or definition:

Cardiovascular	Nervous
Digestive	Reproductive
Endocrine	Respiratory
Integumentary	Skeletal
Lymphatic	Urinary
Muscular	

_____ 1. Organ system that consists of skin, hair, nails and sweat glands; protects and prevents water loss.

_____ 2. Organ system that consists of the brain, spinal cord, and nerves; detects sensation and controls movements.

_____ 3. Organ system that consists of the lungs; exchanges gases between the blood and the air.

_____ 4. Organ system that consists of the kidneys and urinary bladder; removes waste products from the circulatory system.

_____ 5. Organ system that consists of the mouth, pharynx, esophagus, stomach, and intestines; breaks down and absorbs nutrients.

_____ 6. Organ system that consists of bones and cartilage; protects and supports the body, and produces blood cells.

_____ 7. Organ system that consists of the heart, blood vessels, and blood; transports nutrients, wastes, and gases.

_____ 8. Organ system that consists of glands such as the pituitary and thyroid glands; a major regulatory system.

_____ 9. Organ system that consists of muscles attached to the skeleton; allows body movement, maintains posture, and produces body heat.

Homeostasis

66 *Homeostasis is the existence and maintenance of a relatively constant environment within the body.* 99

Match these terms with the
correct statement or definition:

Negative feedback
Positive feedback

_____ 1. Maintains homeostasis by making smaller or resisting any deviation
from a normal value.

_____ 2. When a deviation from a normal value occurs, the response is to
make the deviation even greater.

_____ 3. Medical therapy overcomes illness by aiding this type of feedback.

Directional Terms

66 *Directional terms refer to the body in the anatomical position.* 99

Match these terms with the
correct statement or definition:

Anterior	Medial
Deep	Posterior
Distal	Proximal
Dorsal	Superficial
Inferior	Superior
Lateral	Ventral

_____ 1. Structure below another.

_____ 2. Toward the back of the body (two terms).

_____

_____ 3 Farther from the point of attachment to the body.

_____ 4. Away from the midline of the body.

_____ 5. Away from the surface.

Planes

66 *A plane is an imaginary flat surface passing through the body or an organ.* 99

A. Match these terms with the
correct statement or definition:

| Frontal plane | Sagittal plane |
| Sagittal plane | Transverse plane |

_____ 1. Runs vertically through the body and separates it into equal right
and left portions.

_____ 2. Runs horizontally through the body and divides it into superior and
inferior parts.

_____ 3. Runs vertically through the body and divides it into anterior and
posterior portions.

B. Match these terms with the correct planes labeled in Figure 1-1:

Frontal plane
Midsagittal plane
Transverse plane

1. _____

2. _____

3. _____

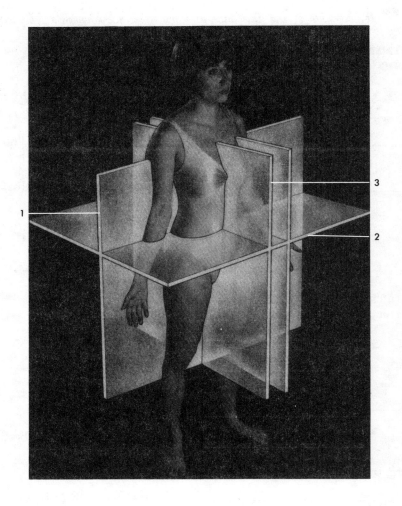

3

1

2

Body Regions

66*The body is commonly divided into several regions.***99**

Using the terms provided, complete the following statements.

Abdomen   Quadrants
Arm   Pelvis
Leg   Regions

The upper limb between the shoulder and elbow is the _(1)_, whereas the lower limb between the knee and the ankle is the _(2)_. The _(3)_ is the inferior end of the trunk associated with the hips, and the region between the thorax and the pelvis is the _(4)_. The abdomen can be divided superficially into four areas called _(5)_, which are used as reference points for locating the underlying organs.

1. _____

2. _____

3. _____

4. _____

5. _____

Body Cavities

The body contains several large trunk cavities that do not open to the outside of the body.

A. **M**atch these terms with the correct statement or definition:

Abdominal cavity
Pelvic cavity
Thoracic cavity

_____ 1. Cavity surrounded by the rib cage and bounded inferiorly by the diaphragm.

_____ 2. Cavity divided into two parts by the mediastinum.

_____ 3. Cavity surrounded by the abdominal muscles and bounded superiorly by the diaphragm.

_____ 4. Small space enclosed by the bones of the pelvis.

_____ 5. Cavity containing the heart and lungs.

_____ 6. Cavity containing the stomach and kidneys.

_____ 7. Cavity containing the urinary bladder and internal reproductive organs.

 There is no physical separation between the abdominal and pelvic cavities, which are sometimes called the abdominopelvic cavity.

B. **M**atch these terms with the correct part labeled in Figure 1-2:

Abdominal cavity
Diaphragm
Mediastinum
Pelvic cavity
Thoracic cavity

1. _____

2. _____

3. _____

4. _____

5. _____

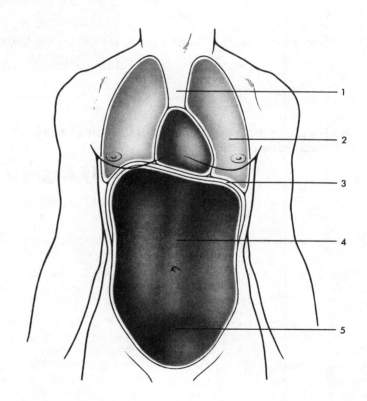

1. Arrange the seven structural levels of the body in order, from the smallest to the largest.

2. List the four primary tissue types.

3. List the two kinds of feedback mechanisms found in living organisms.

4. Describe the anatomical position.

5. List the three major planes used to section the human body.

6. Name the three trunk cavities of the human body. Give an example of an organ found in each trunk cavity.

Place the letter corresponding to the correct answer in the space provided.

_____ 1. Physiology
    a. deals with the processes or functions of living things.
    b. is the scientific discipline that investigates the body's structure.
    c. is concerned with the form of structures and their microscopic organization.
    d. recognizes the unchanging nature of living things.

_____ 2. An organ is
    a. a specialized structure within a cell that carries out a specific function.
    b. at a lower level of organization than a cell.
    c. two or more tissues that perform a specific function.
    d. a group of cells that perform a specific function.

_____ 3. The systems that are most important in the regulation or control of the other systems of the body are the
    a. circulatory and muscular systems.
    b. circulatory and endocrine systems.
    c. nervous and muscular systems.
    d. nervous and endocrine systems.

_____ 4. Negative-feedback mechanisms
    a. make deviations from normal smaller.
    b. maintain homeostasis.
    c. are responsible for an increased rate of sweating when air temperature is higher than body temperature.
    d. all of the above

_____ 5. Body temperatures were measured during an experiment. Results are presented in the graph at the top of the right-hand column. At point A, the subject moved from a swimming pool containing cool water into a jacuzzi containing hot water. As a result, body temperature increased to point B.

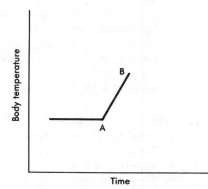

Graphed below are two possible responses to the increase in body temperature.

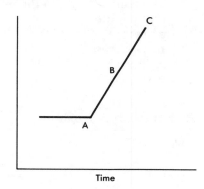

Response 1

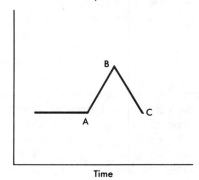

Response 2

Which of the responses graphed above represents a positive-feedback mechanism?
a. Response 1
b. Response 2

_____ 6. Which of the following statements concerning homeostasis is true?
   a. Normal blood pressure is required to ensure that tissue homeostasis is maintained.
   b. If blood pressure decreases below normal, negative-feedback mechanisms increase heart rate.
   c. Low blood pressure that results in inadequate pumping of blood by the heart is an example of positive feedback.
   d. all of the above

_____ 7. Which of the following terms mean the same thing when referring to a human in the anatomical position?
   a. superior and dorsal
   b. deep and distal
   c. anterior and ventral
   d. proximal and medial

_____ 8. The chin is _____ to the umbilicus (belly button).
   a. lateral
   b. posterior
   c. distal
   d. superior

_____ 9. A plane that divides the body into anterior and posterior portions is a
   a. frontal plane.
   b. sagittal plane.
   c. transverse plane.

_____ 10. Which of the following terms is correctly defined?
   a. The arm is that part of the upper limb between the shoulder and wrist.
   b. The leg is that part of the lower limb between the knee and ankle.
   c. The abdomen is most inferior portion of the trunk.
   d. The mediastinum divides the abdominal cavity into left and right parts.

_____ 11. The pelvic cavity contains the
   a. kidneys.
   b. liver.
   c. stomach.
   d. spleen.
   e. urinary bladder.

_____ 12. The thoracic cavity is separated from the abdominal cavity by the
   a. diaphragm.
   b. mediastinum.
   c. mesentery.
   d. rib cage.

☆———————| **FINAL CHALLENGES** |———————☆

Use a separate sheet of paper to complete this section.

1. Complete the following statements, using the correct directional term for a human being.
   a. The knee is _____ to the ankle.
   b. The ear is _____ to the nose.
   c. The nose is _____ to the lips.
   d. The lips are _____ to the front teeth.
   e. The ribs are _____ to the skin.

2. When blood sugar levels decrease, the hunger center in the brain is stimulated. Is this part of a negative or positive feedback system? Explain.

3. When food enters the small intestine, bile is released from the liver into the small intestine. Bile contains bile salts which promote the digestion of fats. Some of the bile salts are reabsorbed from the small intestine into the blood. Theses bile salts circulate to the liver in which they stimulate the release of additional bile containing bile salts. Is this process of bile release an example of negative or positive feedback? Explain.

4. In which quadrants is the pelvic cavity and the heart located? In which quadrant is the liver and the stomach mostly located?

ANSWERS TO CHAPTER 1

Anatomy and Physiology
1. Anatomy; 2. Anatomy; 3. Physiology; 4. Physiology

Structure and Functional Organization
A. 1. Organelle; 2. Cell; 3. Tissue; 4. Organ system
B. 1. Integumentary; 2. Nervous; 3. Respiratory;
4. Urinary; 5. Digestive; 6. Skeletal;
7. Cardiovascular; 8. Endocrine; 9. Muscular

Homeostasis
1. Negative feedback; 2. Positive feedback;
3. Negative feedback

Directional Terms
1. Inferior; 2. Posterior and dorsal; 3. Distal;
4. Lateral; 5. Deep

Planes
A. 1. Midsagittal plane; 2. Transverse plane;
3. Frontal plane
B. 1. Frontal plane; 2. Transverse plane;
3. Midsagittal plane

Body Regions
1. Arm; 2. Leg; 3. Pelvis; 4. Abdomen; 5. Quadrants

Body Cavities
A. 1. Thoracic cavity; 2. Thoracic cavity;
3. Abdominal cavity; 4. Pelvic cavity;
5. Thoracic cavity; 6. Abdominal cavity;
7. Pelvic cavity
B. 1. Mediastinum; 2. Thoracic cavity;
3. Diaphragm; 4. Abdominal cavity; 5. Pelvic
cavity

1. Chemical, organelle, cell, tissue, organ, organ system, organism
2. Epithelial, connective, muscular, and nervous tissues
3. Negative and positive feedback
4. A person standing erect with the feet pointing forward, arms hanging to the sides, and the palms of the hands facing forward
5. Sagittal, transverse, and frontal
6. The thoracic cavity contains the heart, lungs, esophagus, and trachea; the abdominal cavity contains the stomach, liver, spleen, pancreas, kidneys, and most of the intestine; the pelvic cavity contains the urinary bladder and the internal reproductive organs.

1. A. Physiology deals with the processes or functions of living things. Physiology emphasizes the changing nature of living things. Anatomy investigates the body's structure and microscopic organization.

2. C. An organ is two or more tissues that perform a specific function. An organelle is a specialized structure within a cell. Organelles are at a lower level of organization than a cell, but organs are at a higher level. A tissue is a group of cells that perform a specific function.

3. D. The nervous and endocrine systems are the most important regulatory systems of the body. The circulatory system transports gases, nutrients, and waste products. The muscular system is responsible for movement.

4. D. Negative-feedback mechanisms maintain homeostasis by making deviations from normal smaller. When air temperature is greater than body temperature, body temperature tends to increase. Sweating decreases body temperature (reduces deviation from normal) and helps to maintain homeostasis. It is a negative-feedback mechanism.

5. A. First, you must be able to interpret the graphs. For response 1, body temperature increased still further from the normal value, and for response 2 body temperature returned to normal. Next, the definitions of positive and negative feedback must be applied to the graphs. Because positive-feedback mechanisms increase the difference between a value and its normal level (homeostasis), response 1 is a positive-feedback mechanism. Negative-feedback mechanisms resist further change or return the values to normal as in response 2.

6. D. Normal blood pressure ensures adequate delivery of blood to tissues. Consequently the homeostasis of cells is maintained because they receive needed nutrients and gases, and waste products are removed. A decrease in blood pressure is prevented by an increase in heart rate which increases blood pressure back toward normal values. If blood pressure is so low that heart tissue does not receive enough blood, then positive feedback can cause decreased pumping of blood by the heart.

7. C. Anterior (toward the front of the body) and ventral (toward the belly) can be used interchangeably.

8. D. The chin is superior to (higher than) the umbilicus.

9. A. A frontal plane divides the body into anterior and posterior portions. A sagittal plane divides the body into left and right portions, and a transverse plane divides the body into superior and inferior portions.

10. B. The leg is that part of the lower limb between the knee and the ankle. The arm is that part of the upper limb between the shoulder and the elbow. The pelvis is the most inferior portion of the trunk, and the mediastinum divides the thoracic cavity into left and right parts.

11. E. The pelvic cavity contains the urinary bladder, the internal reproductive organs, and the lower part of the digestive tract. The other organs listed are found in the abdominal cavity.

12. A. The diaphragm separates the thoracic cavity from the abdominal cavity. The mediastinum divides the thoracic cavity into left and right parts.

 ☆ **FINAL CHALLENGES** ☆

1. a. Proximal (superior)
   b. Lateral
   c. Superior
   d. Anterior
   e. Deep

2. It is part of a negative-feedback system. Stimulation of the hunger center can result in eating, and the ingested food causes blood sugar levels to increase (return to homeostasis).

3. Bile secretion is an example of positive feedback. As more bile is secreted, more bile salts are reabsorbed, stimulating the secretion of even more bile.

4. The pelvic cavity is in the inferior portion of the right lower and left lower quadrants. The heart, which is in the thoracic cavity, is not located under any of the quadrants, which subdivide the abdominopelvic cavity. The liver is mostly in the right upper quadrant, and the stomach is mostly in the left upper quadrant.

The Chemistry of Life

FOCUS: Chemistry is the study of the composition and structure of substances and the reactions they undergo. Matter is composed of atoms which consist of a nucleus (protons and neutrons) surrounded by electrons. Molecules are formed when two or more atoms are combined by means of chemical bonds. A chemical reaction is the process by which atoms or molecules interact to form or break chemical bonds. Hydrogen ion concentration of a solution, which is measured by pH, affects the chemical behavior of many molecules. Water has many useful properties for living organisms. Important large organic molecules in humans are carbohydrates, lipids, proteins, and nucleic acids.

CONTENT LEARNING ACTIVITY

Basic Chemistry and the Structure of Atoms

66Chemistry is the scientific study of the composition and structure of substances and99 the many reactions they undergo.

A. Match these terms with the correct statement or definition:

Atom   Nucleus
Electron   Neutron
Element   Orbitals
Matter   Proton

_____ 1. Anything that occupies space.

_____ 2. Matter composed of atoms of only one kind.

_____ 3. Smallest particle of an element that retains the properties of that element.

_____ 4. Subatomic particle with no electrical charge; part of the nucleus.

_____ 5. Regions in which the electrons move around the nucleus.

Different elements have different numbers of protons.

B. Match these terms with the correct parts labeled in Figure 2-1:

Electron   Orbital
Nucleus   Proton
Neutron

1. _____

2. _____

3. _____

4. _____

5. _____

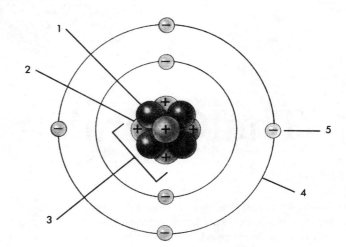

Electrons and Chemical Bonds

66 *Chemical bonds are formed when the outermost electrons are transferred or shared between atoms.* 99

Using the terms provided, complete the following statements:

Carbon   Hydrogen
Covalent   Ionic
Electrolytes   Ions
Electrons   Oxygen

Much of an atom's chemical behavior is determined by the (1) in its outermost orbitals. (2) bonds result when one atom loses an electron and another accepts that electron. Charged atoms that have donated or accepted electrons are called (3). Substances that produce ions when dissolved are sometimes referred to as (4) because ions can conduct electrical current when they are in solution. (5) bonds result when two atoms share one or more pairs of electrons. Hydrogen atoms can share electrons with a(n) (6) atom to form a water molecule. A(n) (7) atom can share four of its electrons with other atoms, forming four covalent bonds.

1. _____

2. _____

3. _____

4. _____

5. _____

6. _____

7. _____

 A combination of atoms held together with chemical bonds is called a molecule; a molecule with two or more different kinds of atoms may be referred to as a compound.

12

Chemical Reactions

" A chemical reaction is the process by which atoms or molecules interact to form or break " *chemical bonds.*

A. Using the terms provided, complete the following statements:

Energy	Reactants
Heat	Released
Molecules	Stored

The atoms or molecules present before the chemical reaction occurs are the (1) and those produced by the chemical reaction are the products. Chemical reactions are important because of the products they form and the changes in (2) they produce. Energy exists in chemical bonds as (3) energy. If products of a chemical reaction contain less stored energy than the reactants, then energy is (4). Although most of the energy is released as (5), the rest is used to form new (6) or to drive processes such as muscle contraction.

1. _____
2. _____
3. _____
4. _____
5. _____
6. _____

B. Using the terms provided, complete the following statements:

ADP	Phosphate
ATP	Released
Currency	Stored

An example of a reaction that releases energy is the breakdown of ATP to (1) and a phosphate group. The phospate group is attached to the ADP molecule by a phosphate bond, which represents (2) energy. When the bond between ADP and the phospate group is broken, energy is (3). Energy released from the breakdown of food molecules is used to form a (4) bond that attaches a phosphate group to ADP. Because ATP molecules can store and provide energy, they are called the energy (5) of the cell. Almost all of the chemical reactions of the cell that require energy use (6) as the energy source.

1. _____
2. _____
3. _____
4. _____
5. _____
6. _____

Acids, Bases, and pH

" The chemical behavior of many molecules changes as the pH of the solution in which they are " *dissolved changes.*

A. Match these terms with the correct statement or definition:

Acids
Bases

Buffers

_____ 1. Substances that are proton (H^+) donors.

_____ 2. Substances that accept protons.

_____ 3. Chemicals that resist changes in pH when acids or bases are added to a solution.

B. Match these terms with the
correct statement or definition:

Acidic solution   Neutral solution
Alkaline (basic) solution

_____ 1. pH of 7 (e.g., pure water).

_____ 2. pH less than 7.

_____ 3. Greater concentration of hydroxide ions than hydrogen ions.

Understanding pH

"_The symbol pH stands for the power (p) of hydrogen ion (H+) concentration._**"**

Using the terms provided, complete the following statements:

1. _____

Acidosis   Hydrogen ions (H$^+$)   2. _____
Alkalosis   7
Carbon dioxide   10   3. _____
Enzymes
                                              4. _____
A change in the pH of a solution by one pH unit represents an
change in _(1)_ by a factor of _(2)_. For example, a solution with   5. _____
a pH of 6 has 10 times the number of hydrogen ions as a solution
with a pH of _(3)_. If pH deviates from its normal range, _(4)_ do   6. _____
not function as well, and may become nonfunctional, harming
normal cell function. The condition of _(5)_ results if the blood   7. _____
pH drops below 7.35; this depresses the nervous system, leading
to disorientation and possible coma. _(6)_ results if blood pH
rises above 7.45; this causes the nervous system to be over-
excitable and may lead to convulsions. The respiratory system
regulates blood pH by controlling _(7)_ levels in the blood

Water

"_A molecule of water consists of one atom of oxygen joined by covalent bonds to two atoms of hydrogen._**"**

Using the terms provided, complete the following statements:

1. _____

Dissociate   Lubricant   2. _____
Digestion   React
Heat   3. _____

Water has many useful properties for living organisms. When   4. _____
ionic substances dissolve in water, the positive and negative
ions separate or _(1)_, allowing the ions to _(2)_ with other   5. _____
molecules. Water is necessary in many chemical reactions, such
as the _(3)_ of food. Water also acts as a _(4)_ in the form of
tears. Water can absorb large amounts of _(5)_ and remain at a
stable temperature, and evaporation of sweat removes heat
from the body.

14

Organic Molecules

"_Organic molecules are those that contain carbon, with the exception of carbon dioxide._**"**

A. Match these terms with the correct statement or definition:

Amino acids   Monosaccharides
Glycerol and fatty acids   Nucleotides

_____ 1. Building blocks of carbohydrates.

_____ 2. Building blocks of fats.

_____ 3. Building blocks of proteins.

_____ 4. Building blocks of nucleic acids.

B. Match these terms with the correct statement or definition:

Disaccharide   Polysaccharide
DNA and RNA   Triglyceride

_____ 1. Two monosaccharides joined together, e. g., sucrose.

_____ 2. Many monosaccharides bound in long chains; examples are glycogen and plant starch.

_____ 3. Most common type of fat molecule.

_____ 4. Important types of nucleic acids.

C. Match these terms with the correct function:

DNA   Glucose and fructose
Fats   RNA

_____ 1. Important carbohydrate energy sources for many of the body's cells.

_____ 2. Functions include energy storage, padding, and insulation.

_____ 3. Hereditary material of the cell; responsible for controlling cell activities.

_____ 4. Three different types are involved in protein synthesis.

D. Using the terms provided, complete the following statements:

Contraction   Framework
Denaturation   Reactants
Enzymes   Shape
Essential

The building blocks of proteins are 20 basic types of amino acids. Humans can synthesize 12 of these from simple organic molecules, but the remaining eight are called _(1)_, and must be included in the diet. Proteins perform many important functions. Structural proteins provide the _(2)_ for many of the body's tissues, muscles contain proteins that are responsible for muscle _(3)_, and _(4)_ are proteins that increase the rate of chemical reactions. The ability of an enzyme to perform its normal function depends on its _(5)_. According to the lock and key model, the enzyme and _(6)_ must fit together, and therefore enzymes are very specific for the reactions they control. _(7)_ occurs when the bonds that maintain a protein's shape are broken, and the protein becomes nonfunctional.

1. _____
2. _____
3. _____
4. _____
5. _____
6. _____
7. _____

Clinical Applications

❝ *Protons, neutrons, and electrons of atoms have properties that can be useful in a clinical setting.* **❞**

Match these terms with the correct statement or definition:

CAT scan   Radioactive
Isotopes   X-ray
MRI

_____

_____

_____

_____

_____

1. Forms of an element with the same number of protons and electrons, but different numbers of neutrons.

2. Isotopes with unstable nuclei that lose neutrons or protons.

3. Radiation formed when electrons lose energy by moving from a higher to a lower energy orbital; used to examine bones and teeth.

4. X-ray "slices" assembled by a computer to produce a 3-D image.

5. Examination method in which radiowaves given off by hydrogen nuclei are used by a computer to produce an image of the body.

QUICK RECALL

1. List three subatomic particles and give their charge.

2. List two types of bonds between atoms.

3. List the pH range of acidic solutions, alkaline solutions, and neutral solutions.

4. List four important properties of water for living organisms.

5. Name the four types of large organic molecules found in living things. For each type of organic molecule, list its building block.

6. List three types of carbohydrate molecules.

7. List three functions of lipids in the human body.

8. List five functions of proteins in the human body.

9. List three functions of nucleic acids in the human body.

MASTERY LEARNING ACTIVITY

Place the letter corresponding to the correct answer in the space provided.

_____ 1. The smallest particles of an element that retain the properties of that element are
   a. electrons.
   b. molecules.
   c. neutrons.
   d. protons.
   e. atoms.

_____ 2. Which of the following statements concerning protons in an atom is correct?
   a. Proton number is the same for all elements.
   b. Proton number is equal to the number of electrons.
   c. Protons are found in regions called orbitals.
   d. Protons have no charge.

_____ 3. A covalent bond occurs when
   a. two atoms share electrons.
   b. an electron is lost from one atom and accepted by another.
   c. an atom becomes ionized.
   d. b and c

_____ 4. Much of an atom's chemical behavior is determined by
   a. the number of neutrons it has.
   b. the electrons in its outermost orbitals.
   c. the total number of neutrons and protons in the nucleus.
   d. the number of electrons in the nucleus.

_____ 5. Most of the chemical reactions of the cell that require energy use _____ as the energy source.
   a. heat
   b. ADP
   c. HCl
   d. CO_2
   e. ATP

_____ 6. A solution with a pH of 5 is a (an) ____ and contains ____ hydrogen ions than a neutral solution.
   a. base, more
   b. base, fewer
   c. acid, more
   d. acid, fewer

_____ 7. A buffer
   a. slows down chemical reactions.
   b. speeds up chemical reactions.
   c. increases the pH of solutions.
   d. maintains a relatively constant pH.

_____ 8. The respiratory system regulates the pH of the blood by controlling the level of
   a. carbon dioxide in the blood.
   b. HCl in the stomach.
   c. water in the blood.
   d. enzymes in the blood.
   e. heat in the body.

_____ 9. Water
   a. is composed of two oxygen atoms and one hydrogen atom.
   b. carries small amounts of heat from the body when it evaporates.
   c. is composed of molecules into which ionic substances dissociate.
   d. is involved in very few chemical reactions in the body.

_____ 10. Which of the following is an example of a carbohydrate?
   a. glycogen
   b. triglyceride
   c. steroid
   d. DNA
   e. none of the above

_____ 11. The building blocks of fats are
   a. simple sugars (monosaccharides).
   b. double sugars (disaccharides).
   c. amino acids.
   d. glycerol and fatty acids.
   e. nucleotides.

_____12. Which of the following molecules is (are) correctly matched with the function?
a. glucose and fructose - energy sources for many cells.
b. fats - energy storage molecules.
c. proteins - structural molecules that provide framework for many tissues.
d. glycogen - storage carbohydrate containing many glucose molecules.
e. all of the above

_____13. A protein
a. is composed of amino acid building blocks.
b. has a function that depends on its shape
c. can be denatured
d. all of the above

_____14. Enzymes
a. function by bringing reactants together.
b. are protein molecules.
c. are very specific for the reactions they control.
d. all of the above

_____15. DNA
a. is the genetic material.
b. is composed of amino acids.
c. contains the sugar ribose.
d. is involved in protein synthesis.
e. all of the above

FINAL CHALLENGES

Use a separate sheet of paper to complete this section.

1. You fill a glass with water, place a teaspoon of salt (NaCl) in it, and note that the salt "disappears". You then let the glass sit until the water evaporates, and the salt "reappears". Explain the apparent disappearance and reappearance of the salt molecules.

2. Two substances, A and B, can combine to form substance C:

$$A + B \rightarrow C$$

Substance A and B each dissolve in water to form a colorless solution, whereas substance C is a red solution. Using this information, explain the following experiment.

When solution A and B are combined, no color change takes place. However, when substance D is added, the combined solution turns red. Later, the exact amount of substance D that was added is recovered from the solution.

3. Given that blood is buffered by the following reaction:

$$CO_2 + H_2O \leftrightarrow H^+ + HCO_3^-$$

What will happen to blood pH if a person holds his (her) breath?

4. Professor Eatwell, in a full-page advertisement, claims to have invented a pill that allows 100% of the energy in the food a person consumes to be converted into usable energy. Even if the claim is true, why might this <u>not</u> be a good idea?

ANSWERS TO CHAPTER 2

Basic Chemistry and the Structure of Atoms
A. 1. Matter; 2. Element; 3. Atom; 4. Neutron;
5. Orbitals
B. 1. Neutron; 2. Proton; 3. Nucleus; 4. Orbital;
5. Electron

Electrons and Chemical Bonds
1. Electrons; 2. Ionic; 3. Ions; 4. Electrolytes;
5. Covalent; 6. Oxygen; 7. Carbon

Chemical Reactions
A. 1. Reactants; 2. Energy; 3. Stored;
4. Released; 5. Heat; 6. Molecules
B. 1. ADP; 2. Stored; 3. Released; 4. Phosphate;
5. Currency; 6. ATP

Acids, Bases, and pH
A. 1. Acids; 2. Bases; 3. Buffers
B. 1. Neutral solution; 2. Acidic solution;
3. Alkaline (basic) solution

Understanding pH
1. Hydrogen ions (H^+); 2. 10; 3. 7; 4. Enzymes;
5. Acidosis; 6. Alkalosis; 7. Carbon dioxide

Water
1. Dissociate; 2. React; 3. Digestion; 4. Lubricant;
5. Heat

Organic Molecules
A. 1. Monosaccharides; 2. Glycerol and fatty acids;
3. Amino acids; 4. Nucleotides
B. 1. Disaccharide; 2. Polysaccharide;
3. Triglyceride; 4. DNA and RNA
C. 1. Glucose and fructose; 2. Fats; 3. DNA; 4. RNA
D. 1. Essential; 2. Framework; 3. Contraction;
4. Enzymes; 5. Shape; 6. Reactants;
8. Denaturation

Clinical Applications
1. Isotopes; 2. Radioactive; 3. X-ray; 4. CAT scan;
5. MRI

1. Electrons - negative charge, protons - positive charge, and neutrons - no charge
2. Ionic bonds and covalent bonds
3. Acidic - pH less than 7, alkaline - pH greater than 7, and neutral - pH 7
4. Many substances dissolve in water, water is a reactant in many reactions, water is a lubricant, and water can absorb large amounts of heat
5. Carbohydrates - monosaccharides; fats - glycerol and fatty acids; proteins - amino acids, and nucleic acids - nucleotides

6. Monosaccharides, disaccharides, and polysaccharides
7. Energy, structure (phospholipids, cholesterol), and regulation (hormones)
8. Regulation (enzymes), structure (collagen), energy, contraction, and transport (hemoglobin)
9. Regulation (cell activities), heredity, and protein synthesis

1. E An element consists of atoms of only one kind, and atoms are the smallest particles that retain the properties of that element. A molecule is composed of two or more atoms joined together, and electrons, protons, and neutrons are parts of atoms.

2. B The number of protons in an atom is equal to the number of electrons, which makes the atom electrically neutral. The number of protons is different for each element. Electrons are the subatomic particles found in orbitals around the nucleus, and protons have a positive charge.

3. A A covalent bond occurs when two atoms share electrons. When an atom donates or accepts an electron, it becomes an ion. Ionic bonds occur when ions of opposite charge are attracted to each other.

4. B. An atom's chemical behavior is mostly determined by the electrons in its outermost orbitals, because chemical bonds are formed when electrons in these orbitals are gained, lost, or shared. Neutrons and protons are both found in the nucleus, and do not have a direct influence on the chemical behavior of the atom. Electrons are not found in the nucleus.

5. E. ATP is the energy source for almost all of the chemical reactions of the cell that require energy. When ATP is broken down, ADP and a phosphate group are produced, and energy is released. If the products of a reaction contain less energy than the reactants, most of the energy is released as heat, but heat is not used as an energy source for reactions. HCl is an acid found in the stomach, and CO_2 is carbon dioxide. Neither of these molecules is used as an energy source in the cell.

6. C. On the pH scale, numbers below 7 denote an acid. Because acids donate hydrogen ions to a solution, the more acidic the solution (i.e., lower pH), the more hydrogen ions are present.

7. D. Buffers resist a change in pH. Enzymes speed up reactions.

8. A. The lungs remove carbon dioxide from the blood. Because the following reaction is reversible :

$$CO_2 + H_2O \leftrightarrow H^+ + HCO_3^-$$

If CO_2 is added to or removed from the blood, the number of hydrogen ions changes, and the pH of the blood changes also. The other factors listed are not involved in the control of blood pH.

9. C. Ionic substances tend to separate, or dissociate in water. Water is composed of two hydrogen atoms, and one oxygen atom. Water absorbs large amounts of heat when it changes from the liquid form (evaporates), and is very important for cooling the body. Water is involved in many chemical reactions in the body.

10. A. Glycogen is a polysaccharide that is composed of many glucose (a monosaccharide) molecules. Triglycerides and steroids are lipids, and DNA is a nucleic acid.

11. D. Glycerol and fatty acids combine to form a fat. The most common fats, which have three fatty acids attached to a glycerol molecule, are triglycerides. Monosaccharides and disaccharides are carbohydrates, amino acids are building blocks for proteins, and nucleotides are the building blocks for DNA and RNA.

12. E. All of the molecules and functions are correctly matched.

13. D. Proteins are composed of amino acid building blocks, have functions that depend on their shape, and can be denatured by increased temperature or pH changes.

14. D. Enzymes function by bringing reactants together, are protein molecules, and control one specific reaction.

15. A. DNA has a double strand of nucleotides (double helix) and contains the sugar deoxyribose. Three types of RNA are involved in protein synthesis.

 ☆ **FINAL CHALLENGES** ☆

1. Salt (NaCl) dissociates into sodium and chloride ions in water. These ions are not visible in water, so the salt has "disappeared". When the water evaporates, the sodium and chloride ions are no longer held apart by the water molecules, so once again sodium chloride molecules appear.

2. Substance D is an enzyme that increases the reaction rate by bringing substances A and B together, but substance D is not used up or changed in the reaction.

3. Holding one's breath causes an increase in blood carbon dioxide levels because carbon dioxide is not eliminated by the respiratory system. As a result, more hydrogen ions (H^+) and bicarbonate ions (HCO_3^-) are formed. The increased number of hydrogen ions causes a decrease in blood pH.

4. Our body temperature is maintained by heat that is "lost" from reactants during chemical reactions. Reducing or eliminating this energy loss would make it impossible to produce our normal body temperature, and death would result. In addition, if food intake was not decreased, considerable weight gain would occur.

Cell Structures and Their Functions

FOCUS: The basic unit of the human body is the cell. Chromosomes in the nucleus contain DNA which regulates the activities of the cell by controlling protein synthesis. The cytoplasm contains organelles which are specialized to perform specific functions such as protein synthesis (ribosomes) and ATP production (mitochondria). The plasma membrane regulates the movement of materials into and out of the cell by diffusion, osmosis, filtration, active transport, and vesicle transport.

CONTENT LEARNING ACTIVITY

Cell Structure and Function

❝*The cell is the basic living unit of all organisms.*❞

Match these terms with the correct statement or definition:

Cytoplasm Nucleus
Extracellular Plasma membrane
Intracellular

_____ 1. Contains the cell's genetic material, which controls the cell.

_____ 2. Living material surrounding the nucleus.

_____ 3. Functions of this structure are to enclose the cytoplasm and form a boundary between material inside and outside the cell.

_____ 4. Substances outside of the cell.

_____ 5. Substances inside the cell.

Plasma Membrane

" *The plasma membrane is a selective barrier that determines what moves into and out of cells.* **"**

A. Using the terms provided, complete the following statements:

Cholesterol   Protein
Phospholipid

The plasma membrane consists of a double layer of _(1)_ molecules and other lipids such as _(2)_ . _(3)_ molecules are on the inner and the outer surfaces of the plasma membrane or extend across the plasma membrane. The protein molecules function as membrane channels, carrier molecules, receptor molecules, enzymes, or structural supports in the membrane.

1. _____

2. _____

3. _____

B. Match these terms with the correct parts labeled in Figure 3-1:

Cholesterol
Membrane channel protein
Phospholipid
Receptor molecule

1. _____

2. _____

3. _____

4. _____

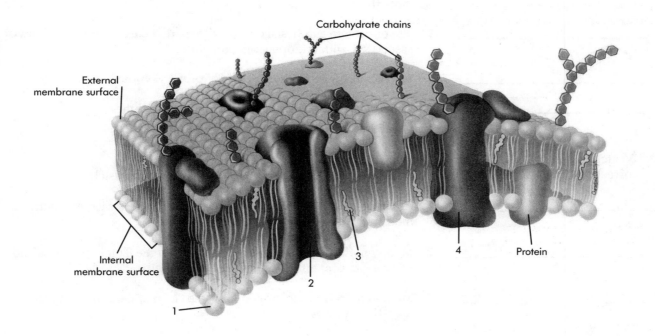

23

Nucleus

"*The nucleus is a large organelle usually located near the center of the cell.***"**

Using the terms provided, complete the following statements:

DNA
Membranes
Nuclear pores

Nucleolus
Ribosomes

1. _____

2. _____

3. _____

4. _____

5. _____

The nucleus contains the cell's genetic material or _(1)_.
Two _(2)_ surround the nucleus. At many points on the surface of
the nucleus, the membranes come together to form _(3)_.
The _(4)_ is a rounded, dense, well-defined nuclear body with no
surrounding membrane, that is the site of manufacture of the
subunits of _(5)_.

☞ Ribosomal subunits leave the nucleus through nuclear pores.

Cytoplasm

"*Cytoplasm is approximately half fluid and half organelles.***"**

A. Match these terms with the
correct statement or definition:

Ribosome
Rough ER
Smooth ER

_____

_____

_____

1. Site of protein synthesis; found as free organelles in the cytoplasm or
associated with endoplasmic reticulum.

2. Membranes with ribosomes; site of protein synthesis

3. Membrane without ribosomes; site of lipid synthesis.

B. Match these terms with the
correct statement or definition:

Golgi apparatus   Lysosome
Mitochondria   Secretory vesicle

_____

_____

_____

_____

1. Closely packed stacks of curved, membrane-bound sacs; concentrates,
modifies, and packages materials.

2. Membrane-bound sac that pinches off from the Golgi apparatus and
releases its contents to the outside of the cell.

3. Membrane-bound sac formed from the Golgi apparatus; contains
digestive enzymes.

4. Rod-shaped organelles with an inner and outer membrane; major site
of ATP production.

C. Match these terms with the correct statement or definition:

Cilia
Flagellum
Microvilli

_____ 1. Projections from the cell surface that move materials along the surface of the cell.

_____ 2. Long projection from the cell surface that functions to move a sperm cell.

_____ 3. Cylindrical extensions of the plasma membrane that do not move; increase the surface area of cells and are important in the absorption of materials.

Cell Diagram

Match these terms with the cell parts labeled in Figure 3-2:

Cilia   Nucleus
Golgi apparatus   Plasma membrane
Lysosome   Ribosome
Microvilli   Rough endoplasmic reticulum
Mitochondrion   Secretory vesicle
Nucleolus   Smooth endoplasmic reticulum

1. _____
2. _____
3. _____
4. _____
5. _____
6. _____
7. _____
8. _____
9. _____
10. _____
11. _____
12. _____

Passive Transport Processes

"_Passive transport processes do not require energy derived from ATP._**"**

Match these terms with the correct statement or definition:

Diffusion	Osmosis
Filtration	Phospholipid layers
Membrane channel	Selectively permeable

_____ 1. General term that means some substances, but not others, can pass through the plasma membrane.

_____ 2. Movement of a substance (other than water) from an area of higher concentration of the substance to an area of lower concentration of the substance.

_____ 3. Protein molecule that allows substances of only a certain size range to pass through.

_____ 4. Water, carbon dioxide, and oxygen pass directly through this part of the plasma membrane.

_____ 5. Movement of water across a selectively permeable membrane from an area of higher water concentration to an area of lower water concentration.

_____ 6. Movement of fluid through a partition containing holes; fluid movement results from the force or weight to the fluid pushing against the partition.

Understanding Osmosis

"_Osmosis is important to cells because large volume changes caused by water movement can disrupt normal cell functions._**"**

Match these terms with the correct statement or definition:

Crenation	Isotonic
Hypotonic	Lysis
Hypertonic	

_____ 1. Solution that has the same concentration of dissolved substances and water as another solution.

_____ 2. When a cell is placed in this type of solution it swells.

_____ 3. Rupture of a cell placed in a hypotonic solution.

_____ 4. When a cell is placed in this type of solution it shrinks.

_____ 5. Term for cell shrinkage.

Active Transport Processes

Active transport processes require the expenditure of energy derived from ATP.

Match these terms with the correct statement or definition:

Active transport
Exocytosis
Phagocytosis

Pinocytosis
Vesicle

_____ 1. Carrier molecules move substances from a lower to a higher concentration.

_____ 2. Membrane-bound droplet found within the cytoplasm of a cell.

_____ 3. Means cell eating and is the movement of solid particles into cells by the formation of a vesicle.

_____ 4. Means cell drinking and is the movement of liquid rather than particles into cells by the formation of small vesicles.

_____ 5. Secretory vesicle fuses with the plasma membrane, and the content of the vesicle is eliminated from the cell.

☞ Cells are able to maintain proper intracellular concentrations of molecules because of the membrane's permeability characteristics and its ability to transport certain molecules.

Protein Synthesis

Events that lead to protein synthesis begin in the nucleus and end in the cytoplasm.

A. Match these terms with the correct statement or definition:

DNA
Gene
mRNA

tRNA
Transcription
Translation

_____ 1. Molecule that directs the production of proteins within cells.

_____ 2. Sequence of nucleotides within a DNA molecule that is a chemical set of instructions for making a specific protein.

_____ 3. This process occurs when the cell makes a copy of the information in DNA necessary to make a specific protein.

_____ 4. Molecule used to carry information from DNA to the ribosomes where the copied information is used to construct a protein.

_____ 5. Transport molecule that carries amino acids to ribosomes.

_____ 6. This process involves the synthesis of proteins at the ribosome.

☞ The proteins produced in a cell function as enzymes or structural components inside and outside the cell.

B. Match these terms with the
structures or processes
labeled in Figure 3-3:

DNA
mRNA
Ribosome

tRNA
Transcription
Translation

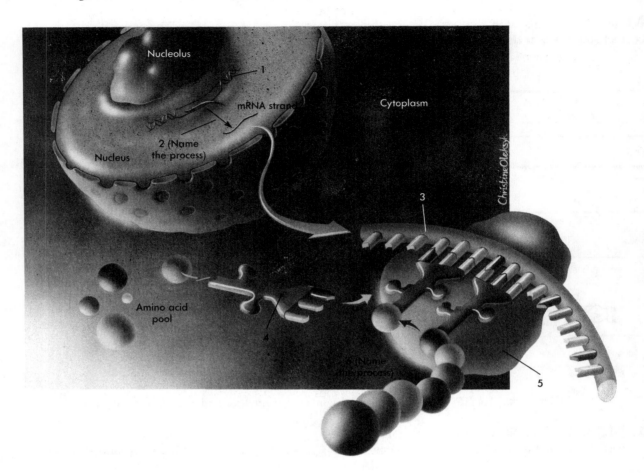

Nucleolus

1

mRNA strand

Cytoplasm

2 (Name
the process)

Nucleus

3

Amino acid
pool

4

Christine Oleksyk

6 (Name
this process)

5

1. _____

3. _____

5. _____

2. _____

4. _____

6. _____

Cell Division and Differentiation

❝*Nearly all cell divisions in the body occur by mitosis, and the resultant "daughter" cells* **❞** *have the same amount and type of DNA as the "parent" cell.*

A. Match these terms with the correct statement or definition:

Chromosome   Interphase
Chromatin   Mitosis

_____ 1. Period between active cell division; time of DNA replication.

_____ 2. Thin strands of DNA spread throughout the nucleus during interphase.

_____ 3. Period during which two sets of genetic material separate into two cells.

_____ 4. Chromatin that has coiled up to form dense bodies.

B. Match these terms with the correct statement or definition:

Anaphase   Prophase
Centriole   Spindle fibers
Metaphase   Telophase

_____ 1. Chromosomes become visible in this phase.

_____ 2. Chromosomes align along the center of the cell.

_____ 3. Chromosomes move toward the ends of the cell.

_____ 4. Nuclear membranes reform.

_____ 5. Organelles that establish the direction in which the chromosomes move.

_____ 6. Moves the chromosomes toward the ends of the cell.

C. Match these terms with the phases of mitosis and the cell parts involved in mitosis labeled in Figure 3-4:

Anaphase
Centriole
Chromosome
Metaphase

Prophase
Spindle fiber
Telophase

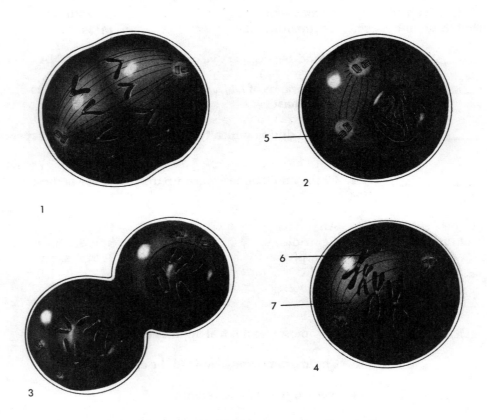

1. _____   4. _____   6. _____

2. _____   5. _____   7. _____

3. _____

D. Using the terms provided, complete the following statements:

The same   Differentiation
Different

All the cells in the body arise from a single fertilized egg. The DNA in these cells is _(1)_. The process by which cells develop specialized structures and functions is called _(2)_. One type of specialized cell (e.g., bone cell) has _(3)_ DNA as another type of specialized cell (e.g., fat cell). The reason specialized cell types are different from each other is because they have _(4)_ parts of their DNA active.

1. _____

2. _____

3. _____

4. _____

Complete the following chart by writing in the organelle described by the structures and functions listed.

ORGANELLE	STRUCTURE	FUNCTION
1. _____	Membrane-bound sac pinched off from the Golgi apparatus.	Contents released to the exterior of the cell by exocytosis.
2. _____	Series of membranes that extend from the outer nuclear membrane into the cytoplasm; no ribosomes attached.	Lipid synthesis.
3. _____	Surrounded by two membranes that join together to form pores.	Contains DNA in the form of chromatin (chromosomes) which produces mRNA.
4. _____	Cylindrical extensions of the plasma membrane; numerous on cells that have them.	Increase cell surface area for absorption.
5. _____	Closely packed stacks of curved membrane-bound sacs.	Concentrates, modifies, and packages materials.
6. _____	Found free in the cytoplasm or associated with endoplasmic reticulum	Site where mRNA and tRNA come together to assemble amino acids into proteins.
7. _____	Membrane-bound vesicle that contains digestive enzymes.	Breakdown of phagocytized particles.
8. _____	Small, bean-shaped or rod-shaped organelle with inner and outer membranes separated by a space.	Most ATP synthesis in the cell.
9. _____	Series of membranes that extend from the outer nuclear membrane into the cytoplasm; ribosomes attached.	Synthesis of proteins.

10. _____

Rounded, dense, well-defined nuclear body.

Manufacture of ribosomal subunits.

11. _____

Projections from the surface of cells that are capable of movement.

Movement of materials along the surface of the cells.

12. _____

Encloses the cytoplasm; composed of phospholipids and proteins

A selective barrier that determines what moves into and out of the cell.

13. List three passive transport processes and two active transport processes that move materials (including water) across the plasma membrane.

14. List three terms used to describe the tendency of cells to shrink or swell when placed in a solution.

15. Name the two steps that occur during protein synthesis.

16. Name in order the four stages of mitosis.

MASTERY LEARNING ACTIVITY

Place the letter corresponding to the correct answer in the space provided.

_____1. Which of the following are functions of the proteins found in cell membranes?
   a. membrane channels
   b. carrier molecules
   c. receptor molecules
   d. enzymes
   e. all of the above

_____2. The nucleolus is
   a. an organelle found in the cytoplasm.
   b. composed of phospholipids.
   c. where ribosomal subunits are manufactured.
   d. where DNA is manufactured.

_____3. The organelle of the cell that serves as the site of protein synthesis?
   a. ribosome
   b. vesicle
   c. Golgi apparatus
   d. microvilli

_____4. The cell organelle that concentrates and packages material to be secreted is the
   a. Golgi apparatus.
   b. nucleolus.
   c. ribosome.
   d. mitochondria.

_____5. The cell organelle that transports materials from the Golgi apparatus to the plasma membrane?
   a. cilia
   b. flagella
   c. mitochondria
   d. secretory vesicle

_____6. Which organelles would one expect to be present in a cell responsible for secreting a lipid?
   a. rough endoplasmic reticulum
   b. smooth endoplasmic reticulum
   c. lysosomes
   d. spindle fibers

_____7. Diffusion
   a. moves substances from a lower to higher concentration.
   b. occurs faster as the concentration difference decreases.
   c. requires the expenditure of energy (ATP).
   d. ends when the diffusing substance is evenly distributed.

_____8. Oxygen and carbon dioxide diffuse through the _____; sodium and potassium ions diffuse through the _____.
   a. membrane channels; membrane channels
   b. membrane channels; phospholipid layers of the plasma membrane
   c. phospholipid layers of the plasma membrane; membrane channels
   d. phospholipid layers of the plasma membrane; phospholipid layers of the plasma membrane

_____9. Cells placed in a hypertonic solution
   a. swell.
   b. shrink.
   c. neither swell nor shrink.

_____10. Filtration
   a. moves water from an area of higher water concentration to an area of lower water concentration.
   b. results from the force or weight of a fluid pushing against a partition containing holes.
   c. requires the expenditure of energy (ATP).
   d. removes large substances, but not small substances, from blood.

_____11. A process that uses vesicles to move liquid (not particulate matter) into cells is
   a. diffusion.
   b. pinocytosis.
   c. phagocytosis.
   d. exocytosis.

33

_____12. In which of the following organelles is mRNA synthesized?
- a. nucleus
- b. ribosome
- c. endoplasmic reticulum
- d. lysosome

_____13. Transfer RNA
- a. is a protein.
- b. carries amino acids to ribosomes.
- c. contains the information necessary to produce a protein.
- d. is also called a gene.

_____14. Given the following activities:
1. repair
2. growth
3. formation of reproductive cells
4. differentiation

Which of the activities are the result of mitosis?
- a. 2
- b. 3
- c. 1, 2
- d. 3, 4
- e. 1, 2, 4

_____15. Given the following events:
1. chromatin coils up to form chromosomes
2. chromosomes move toward opposite ends of the cell
3. chromosomes line up along the center of the cell
4. DNA is duplicated

Arrange the events in the order they occur.
- a. 1, 4, 2, 3
- b. 1, 4, 3, 2
- c. 4, 1, 2, 3
- d. 4, 1, 3, 2

FINAL CHALLENGES

Use a separate sheet of paper to complete this section.

1. If you observed the following characteristics in an electron micrograph of a cell:
1. many microvilli
2. many mitochondria
3. the cell lines a cavity
4. little smooth endoplasmic reticulum
5. little rough endoplasmic reticulum
6. few vesicles

On the basis of these observations, which of the following is likely to be a major function of the cell?
- a. secretion of a protein
- b. secretion of a lipid
- c. intracellular digestion
- d. active transport

2. Mature red blood cells can
- a. synthesize ATP.
- b. divide by mitosis.
- c. synthesize new proteins.
- d. all of the above

3. Container A contains a 10% salt solution and container B a 20% salt solution. If salt and water diffuse between the two solutions, water would move from _____ to _____, and salt would move from _____ to _____.
- a. A, B, A, B
- b. A, B, B, A
- c. B, A, A, B
- d. B, A, B, A

4. Suppose that a woman was running a long distance race in which she lost a large amount of hypotonic sweat. Her cells would
- a. shrink.
- b. swell.
- c. remain the same.

5. Suppose that a man is doing heavy exercise in the hot summer sun. He perspires profusely. He then drinks a large volume of distilled water. His cells would
- a. shrink.
- b. swell.
- c. remain the same.

34

ANSWERS TO CHAPTER 3

CONTENT LEARNING ACTIVITY

Cell Structure and Function
1. Nucleus; 2. Cytoplasm; 3. Plasma membrane;
4. Extracellular; 5. Intracellular

Plasma Membrane
A. 1. Phospholipid; 2. Cholesterol; 3. Protein
B. 1. Phospholipid; 2. Membrane channel protein;
3. Cholesterol; 4. Receptor molecule

Nucleus
1. DNA; 2. Membranes; 3. Nuclear pores;
4. Nucleolus; 5. Ribosomes

Cytoplasm
A. 1. Ribosome; 2. Rough ER; 3. Smooth ER
B. 1. Golgi apparatus; 2. Secretory vesicle;
3. Lysosome; 4. Mitochondria
C. 1. Cilia; 2. Flagellum; 3. Microvilli

Cell Diagram
1. Cilia; 2. Secretory vesicle; 3. Golgi apparatus;
4. Smooth endoplasmic reticulum; 5. Nucleus;
6. Nucleolus; 7. Rough endoplasmic reticulum;
8. Mitochondrion; 9. Plasma membrane;
10. Ribosome; 11. Lysosome; 12. Microvilli

Passive Transport Processes
1. Selectively permeable; 2. Diffusion;
3. Membrane channel; 4. Phospholipid layers;
5. Osmosis; 6. Filtration

Understanding Osmosis
1. Isotonic; 2. Hypotonic; 3. Lysis; 4. Hypertonic;
5. Crenation

Active Transport Processes
1. Active transport; 2. Vesicle; 3. Phagocytosis;
4. Pinocytosis; 5. Exocytosis

Protein Synthesis
A. 1. DNA; 2. Gene; 3. Transcription; 4. mRNA;
5. tRNA; 6. Translation
B. 1. DNA; 2. Transcription; 3. mRNA; 4. tRNA;
5. Ribosome; 6. Translation

Cell Division and Differentiation
A. 1. Interphase; 2. Chromatin; 3. Mitosis;
4. Chromosome
B. 1. Prophase; 2. Metaphase; 3. Anaphase;
4. Telophase; 5. Centrioles; 6. Spindle fibers
C. 1. Anaphase; 2. Prophase; 3. Telophase;
4. Metaphase; 5. Centriole; 6. Chromosome;
7. Spindle fiber
D. 1. The same; 2. Differentiation; 3. The same;
4. Different

QUICK RECALL

1. Secretory vesicle
2. Smooth endoplasmic reticulum
3. Nucleus
4. Microvilli
5. Golgi apparatus
6. Ribosomes
7. Lysosome
8. Mitochondria
9. Rough endoplasmic reticulum
10. Nucleolus
11. Cilia
12. Plasma membrane
13. Passive transport processes: diffusion, osmosis, and filtration; active transport processes: active transport and vesicle transport
14. Hypotonic (cell swells) isotonic (cell stays the same size), hypertonic (cell shrinks)
15. Transcription and translation.
16. Prophase, metaphase, anaphase, and telophase; interphase, the time between cell divisions, is not part of mitosis

MASTERY LEARNING ACTIVITY

1. E.   Proteins in the plasma membrane function as membrane channels, carrier molecules, receptor molecules, enzymes, and structural supports.

2. C.   The nucleolus is the site of manufacture of ribosomal subunits. The nucleolus is found in the nucleus. Phospholipids are a major component of the plasma membrane. DNA is replicated within the nucleus during interphase.

3. A At the ribosomes, mRNA and tRNA form proteins during the process of translation.

4. A The Golgi apparatus concentrates and packages materials to be secreted.

5. D Secretory vesicles pinch off from the Golgi apparatus, move to the plasma membrane, fuse with the plasma membrane, and release their contents to the exterior of the cell by exocytosis.

6. B Smooth endoplasmic reticulum produces lipids, whereas rough endoplasmic reticulum produces proteins.

7. D Diffusion ends when the diffusing substance is evenly distributed, because diffusion moves a substance from an area of higher to lower concentration. The greater the concentration difference, the faster diffusion takes place. Diffusion is a passive process and does not require ATP.

8. C Oxygen, carbon dioxide, water, and urea can diffuse directly through the phospholipid layers of the plasma membrane. Sodium and potassium ions diffuse through membrane channels.

9. B Cells placed in a hypertonic solution shrink, whereas cells placed in a hypotonic solution swell. Cells in an isotonic solution neither swell nor shrink.

10. B Filtration results from the force or weight of a fluid pushing against a partition containing holes. This process does not require ATP. As a result of the force or weight, fluid and substances smaller than the holes pass through the partition. Substances larger than the holes do not pass through. Osmosis, not filtration, is the movement of water from an area of higher water concentration to an area of lower water concentration.

11. B Pinocytosis uses a vesicle to move liquid into the cell. Phagocytosis moves particulate matter into cells, and exocytosis moves materials out of cells. Diffusion does not use vesicles.

12. A mRNA is produced by transcription in the nucleus.

13. B tRNA is a nucleic acid that carries amino acids to ribosomes. mRNA is a copy of the information in a gene (DNA), which is necessary to make a protein. mRNA is formed during transcription. The mRNA and the tRNA interact at the ribosome during the process of translation to make a protein.

14. C Repair and growth through mitosis results in the production of new daughter cells that have the same structures and functions as the parent cell. All cells of the body, except reproductive cells, are produced by mitosis. Differentiation is the process by which cells become different from each other and have different structures and functions.

15. D DNA is replicated during interphase before mitosis begins. During mitosis, chromatin coils up to form a duplicated set of chromosomes (prophase), the duplicated chromosomes line up along the center of the cell (metaphase), and the duplicated chromosomes separate from each other and an identical set of chromosomes moves toward opposite ends of the cell (anaphase).

 FINAL CHALLENGES

1. D The cell lines a cavity, indicating that substances entering or leaving the cavity must move across the cellular barrier. The microvilli increase the surface area of the cell exposed to the cavity, and numerous mitochondria suggest that ATP is used by the cell for active transport.

   Lipid secretion is unlikely with the small amount of smooth endoplasmic reticulum, and protein secretion is unlikely with the small amount of rough endoplasmic reticulum. Few vesicles also suggest that exocytosis is not occurring.

2. A Mature red blood cells, like other cells, have mitochondria and can synthesize ATP. Mature red blood cells, however, do not have a nucleus. Therefore they do not contain the genes necessary for synthesizing proteins and they are not capable of mitosis.

3. B Because solution A has 10% salt, it has 90% water, whereas solution B has 20% salt and 80% water. Substances diffuse from areas of higher concentration to areas of lower concentration. Thus water diffuses from solution A (90% water) to solution B (80% water), and salt diffuses from solution B (20% salt) to solution A (10% salt).

4. A Hypotonic sweat is less concentrated than blood. Thus compared to blood, more water is lost than salts, and the concentration of the blood increases. As a result cells have more water than blood, water moves from the cells into the blood, and the cells shrink.

5. B Replacing the lost water and salts with distilled water results in a more dilute blood. The blood has more water than the cells, water moves from the blood into the cells, and the cells swell.

Tissues, Glands, and Membranes

FOCUS: The cells of the body are specialized to form four basic types of tissues. Epithelial tissue covers free surfaces of the body or forms glands. Connective tissue joins cells and other tissues together, forms a supporting framework for the body (e.g., bone), and transports substances (e.g., blood). Connective tissue is characterized by large amounts of extracellular matrix that separates cells from each other. Muscle tissue has the ability to contract, making body movement (skeletal muscle), blood movement (cardiac muscle) and movement through hollow organs (smooth muscle) possible. Nervous tissue is specialized for conducting action potentials, or impulses. Membranes are thin sheets or layers of tissue that cover a structure or line a cavity. Mucous membranes line cavities that open to the exterior, whereas serous membranes line trunk cavities that do not open to the outside of the body.

CONTENT LEARNING ACTIVITY

Epithelial Tissue

❝*Epithelium covers surfaces of the body or forms glands.***❞**

Match these terms with the
correct statement or definition:

Basement membrane
Free surface

_____ 1. Part of epithelial cells which is not in contact with other cells.

_____ 2. Attaches epithelial cells to underlying tissue.

 A tissue is a group of cells with similar structure and functions and the material between the cells. Histology is the study of tissues and their function.

Classification

"Epithelia are named according to the number of cell layers and the shape of the cells."

A. Match these terms with the correct statement or definition:

Simple columnar epithelium   Stratified squamous epithelium
Simple cuboidal epithelium   Transitional epithelium
Simple squamous epithelium

_____   1. Single layer of thin, flat cells.

_____   2. Single layer of cube-shaped cells.

_____   3. Single layer of tall, thin cells.

_____   4. Multiple layers of cells in which the basal layer is cuboidal and becomes flattened at the surface.

_____   5. Layers of cells that appear cubelike when an organ is relaxed and flattened when the organ is distended by fluid.

B. Match these terms with the correct parts labeled in Figure 4-1:

Basement membrane   Simple squamous epithelium
Free surface   Transitional epithelium
Simple columnar epithelium

1. _____   3. _____   5. _____

2. _____   4. _____

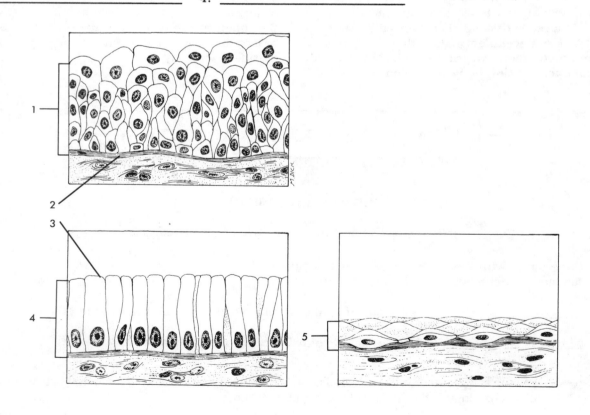

38

C. Match these terms with the
correct statement or definition:

Simple squamous   Simple cuboidal or columnar
Stratified squamous

_____ 1. Epithelium found in organs where the principal function is diffusion,
e.g., alveoli of lungs.

_____ 2. Epithelium found in areas where secretion and absorption occur, e.g.,
the kidneys and small intestine.

_____ 3. Epithelium found in areas where protection is a major function, e.g.,
skin, anus, and vagina.

☞ The greater volume of cuboidal and columnar epithelial cells enables them to contain
organelles responsible for their function.

Glands

66 *A gland is a single cell or a multicellular structure that secretes substances onto a surface, into a* 99
cavity, or into the blood.

A. Match these terms with the
correct statement or definition:

Endocrine
Exocrine

_____ 1. Glands with a duct (e.g., sweat glands).

_____ 2. Glands with no duct that secrete hormones (e.g. thyroid gland).

B. Match these terms with the correct
part labeled in Figure 4-2:

Compound acinar (alveolar) gland
Compound tubular gland
Simple acinar (alveolar) gland
Simple tubular gland
Single-cell gland

1. _____

2. _____

3. _____

4. _____

5. _____

39

Connective Tissue

"Connective tissue functions to hold cells and tissues together, provides a supporting framework for the body, and transports substances."

A. Match these terms with the correct statement or definition:

Adipose
Blood
Bone
Cartilage

Dense connective tissue
Loose (areolar) connective tissue

_____ 1. Extracellular matrix is tightly packed collagen fibers; found in tendons, ligaments, and the dermis.

_____ 2. Extracellular matrix is protein fibers widely separated from each other; provides attachment between organs, around glands, muscles, and nerves, and attaches the skin.

_____ 3. Very little extracellular matrix; filled with lipid for energy storage.

_____ 4. Extracellular matrix contains collagen and traps water; rigid but springs back after being compressed.

_____ 5. Extracellular matrix is mineralized and hard; provides support and protection for other tissues and organs.

_____ 6. Extracellular matrix is liquid; carries materials throughout the body.

☞ The structure of the extracellular matrix determines the functional characteristics of the connective tissue.

B. Match these terms with the correct statement or definition:

Chondrocytes
Fibroblasts

Lacunae
Osteocytes

_____ 1. Produce collagen fibers.

_____ 2. Cartilage cells.

_____ 3. Bone cells.

_____ 4. Spaces containing cells within the matrix of bone or cartilage.

C. Match these terms with the correct parts labeled in Figure 4-3:

Adipose
Bone
Cartilage
Chondrocyte
Dense connective tissue

Fibroblast
Lacuna
Lipids
Osteocyte

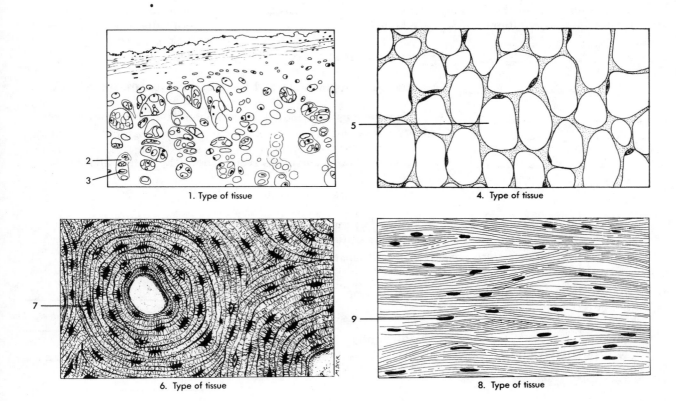

1. Type of tissue

4. Type of tissue

6. Type of tissue

8. Type of tissue

1. _____   4. _____   7. _____

2. _____   5. _____   8. _____

3. _____   6. _____   9. _____

Muscle Tissue

❝*The main characteristic of muscle tissue is its ability to contract or shorten, making movements possible.*❞

Match these terms with the correct statement or definition:

Cardiac muscle
Skeletal muscle

Smooth muscle

_____   1. Cylindrical, striated, voluntary muscle cells with many nuclei.

_____   2. Striated, branching, involuntary cells with intercalated disks and one nucleus.

_____   3. Cells with one nucleus, tapered at each end, unstriated, and involuntary.

41

Nervous Tissue

"*Nervous tissue forms the brain, spinal cord and nerves; it is responsible for coordinating and controlling many of the body's activities.***"**

Match these terms with the correct statement or definition:

Axon   Dendrites
Cell body   Neuroglia

_____   1. Part of the nerve cell containing the nucleus.

_____   2. Receive action potentials and conduct them toward the cell body.

_____   3. A single extension; conducts action potentials away from the cell body.

_____   4. Support cells of the nervous system; nourish, protect, and insulate neurons.

Inflammation

"*The inflammatory response, or inflammation, occurs when tissues are damaged.***"**

A. Using the terms provided, complete the following statements:

Dilation   Neutrophils
Disturbance of function Pain
Edema   Pus
Mediators

Following an injury, chemical substances called _(1)_ of inflammation are released from injured tissues and adjacent blood vessels. Some mediators cause _(2)_ of blood vessels, which causes redness and heat. In this way, materials for repair and fighting infection are brought to the injury site. Mediators of inflammation also increase the permeability of blood vessels, resulting in swelling, or _(3)_ of the tissues. Blood cells called _(4)_ enter the tissues in large numbers, and as they die, dead cells and fluid accumulate as _(5)_. _(6)_ can occur as a result of stimulation of nerve endings from direct damage, chemical mediators of inflammation, or pressure from edema and pus. Edema, pain, and tissue destruction result in _(7)_, which protects the injured area from further damage.

1. _____

2. _____

3. _____

4. _____

5. _____

6. _____

7. _____

B. Match these drugs with the correct effect:

Antihistamine
Aspirin

_____   1. Counteracts the effect of histamines released in hay fever.

_____   2. Prevents synthesis of prostaglandins.

C. Match these terms with the
correct statement or definition:

Tissue regeneration   Tissue replacement
Tissue repair

_____ 1. General term for the substitution of viable cells for dead cells.

_____ 2. Tissue repair in which new cells are the same type as destroyed cells.

_____ 3. Repair in which the tissue formed has cells of a different type than
destroyed cells; scar tissue is formed.

Membranes

66 *A membrane is a thin sheet or layer of tissue that covers a structure or lines a cavity.* 99

A. Match these terms with the
correct statement or definition:

Mucous membrane   Serous membrane
Other membranes

_____ 1. Lines cavities that open to the outside of the body.

_____ 2. Lines the trunk cavities and covers the organs located within the
trunk cavities.

_____ 3. Includes skin, synovial membrane, and periosteum.

B. Match these terms with the
correct statement or definition:

Pleural membranes   Peritoneal membrane
Pericardial membranes

_____ 1. Cover the lungs and line the cavity around the lungs.

_____ 2. Cover the outer surface of the heart and line the cavity surrounding
the heart.

_____ 3. Cover the surface of many abdominopelvic organs and line the
abdominopelvic cavity.

☞ Pericarditis is inflammation of the pericardial membranes, pleurisy is inflammation of the
pleural membranes, and peritonitis is inflammation of the peritoneal membranes.

C. Match these terms with the
correct statement or definition:

Parietal   Visceral

_____ 1. Portion of serous membrane that lines the surface of the body cavity.

_____ 2. Portion of serous membrane that covers the surface of organs found in
the body cavity.

☞ The space between parietal and visceral portions of the serous membranes is filled with serous
fluid, which acts as a lubricant for movement of organs in the body cavities.

Understanding Cancer

"_Oncology is the study of cancer and its associated problems._**"**

A. Match these terms with the correct statement or definition:

| Benign | Metastasis |
| Malignant | Neoplasm |

_____ 1. A tumor; produced by abnormal cell division.

_____ 2. Tumors that enlarge but do not spread to other sites in the body, are surrounded by a connective tissue capsule, and have cells similar to those in normal tissue.

_____ 3. Tumors that enlarge and spread to other sites in the body, are not surrounded by a connective tissue capsule, and can have cells very different from cells of normal tissue.

_____ 4. Movement of tumor cells through the circulatory or lymphatic system to new sites.

B. Match these word endings with the correct definition:

| -carcinoma | -sarcoma |
| -oma | |

_____ 1. Ending for names of most benign tumors.

_____ 2. Ending for names of malignant tumors originating from epithelial tissue.

_____ 3. Ending for names of malignant tumors originating from connective tissue.

C. Using the terms provided, complete the following statements:

Biopsy	Killing
Chemotherapy	Radiation
Imaging	Surgically
Immune	

Anatomical _(1)_ is frequently used to visualize suspected tumors, including x-rays, ultrasound, CAT scans and MRI. Once a tumor is detected, a small piece of tissue is removed in a process called a _(2)_. Cancer therapy is directed at confining and then _(3)_ malignant cells. This is accomplished by killing the tumor with _(4)_ or lasers, removing the tumor _(5)_, by treating the patient with drugs that selectively kill rapidly dividing cells, which is called _(6)_, or by stimulating the patient's _(7)_ system to destroy the tumor.

1. _____

2. _____

3. _____

4. _____

5. _____

6. _____

7. _____

1. List five kinds of epithelium based on numbers of cell layers and shape of cells.

2. List four functions of epithelial cells.

3. List two types of glands, based on whether or not a duct is present.

4. List four types of connective tissue based on the extracellular matrix present.

5. List three types of muscle tissue found in the human body.

6. List the two major categories of membranes found in the human body.

7. Name three serous membranes and the internal organs with which they are associated.

8. List the general terms used for serous membranes that cover the surface of organs and serous membranes that line body cavities.

Place the letter corresponding to the correct answer in the space provided.

_____1. A tissue that covers a surface, is one cell layer thick, and is composed of flat cells is
a. simple squamous epithelium.
b. simple cuboidal epithelium.
c. simple columnar epithelium.
d. stratified squamous epithelium.
e. transitional epithelium.

_____2. Given the following characteristics:
1. capable of contraction
2. covers all free body surfaces
3. lacks blood vessels
4. comprises various glands
5. attached to underlying tissue by a basement membrane.

Which of the above are characteristics of epithelial tissue?
a. 1,2,3
b. 2,3,5
c. 3,4,5
d. 1,2,3,4
e. 2,3,4,5

_____3. Stratified epithelium is usually found in areas of the body where the principal activity is
a. secretion.
b. protection.
c. absorption.
d. diffusion.

_____4. In parts of the body such as the urinary bladder, where considerable stretching occurs, one can expect to find which of the following type of epithelial cells?
a. simple cuboidal
b. simple squamous
c. stratified squamous
d. transitional

_____5. An exocrine gland with many branches, and with the ends of the ducts expanded into a saclike structures is a
a. simple tubular gland.
b. simple acinar (alveolar) gland.
c. compound tubular gland.
d. compound acinar (alveolar) gland.

_____6. The fibers in dense connective tissue are formed by
a. fibroblasts.
b. adipose cells.
c. osteocytes.
d. chondrocytes.

_____7. A tissue that contains a large amount of extracellular collagen organized as closely packed, parallel fibers would probably be found in
a. a muscle.
b. a tendon.
c. adipose tissue.
d. bone.
e. cartilage.

_____8. Which of the following is true of adipose tissue?
a. site of energy storage
b. a type of connective tissue
c. acts as a protective cushion
d. functions as a heat insulator
e. all of the above

_____9. Blood is an example of
a. epithelial tissue.
b. connective tissue.
c. muscle tissue.
d. nervous tissue.
e. none of the above

_____10. Which of the following is characteristic of skeletal muscle?
a. under involuntary (unconscious) control
b. cells tapered at each end
c. intercalated disks present
d. several nuclei per cell
e. all of the above

11. Which of the following statements about nervous tissue is NOT true?
    a. Neurons have cell processes (extensions) called axons.
    b. Electrical signals (action potentials) are conducted along axons.
    c. Dendrites contain the nucleus, and control general cell function.
    d. Neurons are nourished and protected by neuroglia.

12. Chemical mediators of inflammation
    a. stimulate nerve endings to produce pain.
    b. increase the permeability of blood vessels.
    c. cause vasodilation of blood vessels.
    d. are released into or activated in tissues following an injury.
    e. all of the above

13. Which of the following are symptoms of inflammation?
    a. redness and heat
    b. pain
    c. swelling
    d. disturbance of function
    e. all of the above

14. Linings of the digestive, respiratory, excretory, and reproductive tracts are composed of
    a. serous membranes.
    b. synovial membranes.
    c. periosteum.
    d. mucous membranes.

15. The serous membrane covering the outer surface of the lungs is the
    a. parietal pleura.
    b. parietal pericardium.
    c. visceral pleura.
    d. visceral peritoneum.
    e. parietal peritoneum.

16. An osteosarcoma is a
    a. benign tumor of epithelial tissue.
    b. malignant tumor of epithelial tissue.
    c. benign tumor of bone.
    d. malignant tumor of bone.
    e. benign tumor of fibrous connective tissue.

FINAL CHALLENGES

Use a separate sheet of paper to complete this section.

1. On a histology exam, Slide Mann was asked to identify the types of epithelial tissue lining the surface of an organ. He identified the first tissue as stratified squamous epithelium and the second tissue as stratified cuboidal. In both cases he was wrong. Given that the tissues both came from the same organ, what was the epithelial type?

2. A bullet wound that passes through one's upper arm without hitting bone could contact what tissue types?

3. Ronny Beke remembers that an earlier attack of hay fever produced a runny nose, so he treats his cold symptoms by taking his hay fever medication. The hay fever medication is an antihistamine. Would this treatment be helpful? Explain.

4. Han Chu has a cold, which is caused by a viral infection. The virus damages and kills cells in the nasal cavity. Explain why this procedure produces a "stopped up" nose.

5. "Raddy" McDude was riding his skateboard, tried to jump a park bench, and severely twisted his knee. Upon examination, the doctor determined that he had torn cartilage in his knee, and that the torn cartilage must be surgically removed. Why didn't the doctor tell Raddy to just rest the knee until the cartilage healed?

ANSWERS TO CHAPTER 4

CONTENT LEARNING ACTIVITY

Epithelial Tissue
1. Free surface; 2. Basement membrane

Classification
A. 1. Simple squamous epithelium; 2. Simple cuboidal epithelium; 3. Simple columnar epithelium; 4. Stratified squamous epithelium; 5. Transitional epithelium
B. 1. Transitional epithelium; 2. Basement membrane; 3. Free surface; 4. Simple columnar epithelium; 5. Simple squamous epithelium 4. Simple squamous epithelium
C. 1. Simple squamous; 2. Simple cuboidal or columnar; 3. Stratified squamous

Glands
A. 1. Exocrine; 2. Endocrine
B. 1. Single-cell gland; 2. Simple tubular gland; 3. Simple acinar (alveolar) gland; 4. Compound tubular gland; 5. Compound acinar (alveolar) gland

Connective Tissue
A. 1. Dense connective tissue; 2. Loose (areolar) connective tissue; 3. Adipose; 4. Cartilage; 5. Bone; 6. Blood
B. 1. Fibroblasts; 2. Chondrocytes; 3. Osteocytes; 4. Lacunae
C. 1. Cartilage; 2. Lacuna; 3. Chondrocyte; 4. Adipose; 5. Lipids; 6. Bone; 7. Osteocyte; 8. Dense connective tissue; 9. Fibroblast

Muscle Tissue
1. Skeletal muscle; 2. Cardiac muscle; 3. Smooth muscle

Nervous Tissue
1. Cell body; 2. Dendrites; 3. Axon; 4. Neuroglia

Inflammation
A. 1. Mediators; 2. Dilation; 3. Edema; 4. Neutrophils; 5. Pus; 6. Pain; 7. Disturbance of function
B. 1. Antihistamine; 2. Aspirin
C. 1. Tissue repair; 2. Tissue regeneration; 3. Tissue replacement

Membranes
A. 1. Mucous membrane; 2. Serous membrane; 3. Other membranes
B. 1. Pleural membranes; 2. Pericardial membranes; 3. Peritoneal membranes
C. 1. Parietal; 2. Visceral

Cancer
A. 1. Neoplasm; 2. Benign; 3. Malignant; 4. Metastasis
B. 1. -oma; 2. -carcinoma; 3. -sarcoma
C. 1. Imaging; 2. Biopsy; 3. Killing; 4. Radiation; 5. Surgically; 6. Chemotherapy; 7. Immune

QUICK RECALL

1. Simple squamous, simple cuboidal, simple columnar, stratified squamous, and transitional epithelium
2. Diffusion (simple squamous), secretion and absorption (simple cuboidal and simple columnar), protection (stratified squamous), and expansion (transitional)
3. Exocrine and endocrine
4. Matrix with collagen fibers - dense connective tissue and loose or areolar connective tissue; little extracellular matrix - adipose tissue; matrix with collagen and trapped water - cartilage; mineralized matrix - bone; and liquid matrix - blood
5. Skeletal, cardiac, and smooth muscle
6. Mucous and serous membranes
7. Pleural membranes - lungs; pericardial membranes - heart; peritoneal membranes - most abdominopelvic organs
8. Organ surface - visceral; body cavity - parietal

1. A   The tissue is simple (one cell layer thick), squamous (flat cells) epithelium (covers a surface).

2. E   Muscle tissue, not epithelial tissue, is capable of contraction. The other statements are true for epithelial tissue.

3. B   Secretion absorption, and diffusion are functions of simple epithelium. Protection is more often a function of stratified epithelium. As outer cell layers are damaged, they can be replaced by deeper layers.

4. D   Transitional epithelium is composed of cells that can flatten and slide over each other. This allows stretching of the tissue and makes transitional epithelium an ideal lining for the urinary bladder.

5. D   The gland would be a compound (many branches) acinar or alveolar (end of ducts expanded into a saclike structure).

6. A   Fibroblasts secrete the fibers of connective tissue. Adipose cells are cells filled with lipids. Osteocytes are cells located in the lacunae of bone, whereas chondrocytes are cells located in the lacunae of cartilage.

7. B   The characteristics are consistent with dense connective tissue, which is the type found in tendons and ligaments. Parallel collagen fibers give strength in one direction; this is a characteristic necessary for tendons and ligaments, which make muscle to bone and bone to bone connections.

8. E   Adipose tissue is a type of connective tissue. It acts as a protective cushion (around the kidneys, for example) and acts as a heat insulator under the skin.

9. B   You simply had to know that blood is an example of connective tissue.

10. D   Skeletal muscle cells are long, cylindrical, striated, have many nuclei per cell, and are under conscious control. Unlike cardiac muscle, they have no intercalated disks.

11. C   Dendrites are extensions from the nerve cell that receive action potentials and conduct them toward the cell body. The cell body is the location of the nucleus of the cell. Axons conduct action potentials away from the cell body.

12. E   Chemical mediators, such as histamine and prostaglandins are released or activated in tissue following injury to the tisue. The mediators cause vasodilation, increase vascular permeability, and stimulate neurons to produce pain.

13. E   All the symptoms listed are characteristic of inflammation.

14. D   Cavities that open to the exterior are lined by mucous membranes. Serous membranes line the trunk cavities that do not open to the exterior. The cutaneous membrane (skin) covers the outside of the body, synovial membranes line joints, and periosteum covers bones.

15. C   The serous membrane that covers the surface of the lungs is the visceral pleura. Parietal pleura lines the body cavity around the lungs, pericardial membranes surround the heart, and peritoneal membranes are found in the abdominopelvic cavity.

16. D   The prefix "osteo-" indicates bone, and "sarcoma" indicates a malignant tumor originating from connective tissue. If the tumor originates from epithelial tissue and is malignant, the name usually ends with "carcinoma". If the tumor is benign, the name ends with "oma".

1. The tissue is transitional epithelium, a stratified epithelium that lines organs such as the urinary bladder and ureters. When the organ is stretched, the cells of transitional epithelium become squamouslike; and when the organ is not stretched, the cells are roughly cuboidal in shape.

2. A bullet that passes through the upper arm without hitting bone could pass through epithelium (part of the skin), connective tissue (part of the skin and material around muscle), muscle, and nervous tissue (particularly if pain is a result).

3. Chemical mediators of inflammation normally produce beneficial responses, such as dilation of blood vessels and increased blood vessel permeability. Blocking these effects could reduce the ability of the body to deal with harmful agents such as bacteria. On the other hand, antihistamines also reduce many of the unpleasant symptoms of inflammation, making the patient more comfortable and this can be considered beneficial. Antihistamines are commonly taken to prevent allergy symptoms that are often an overreaction of the inflammatory response to foreign substances such as pollen.

4. The viral infection has stimulated an inflammatory response. Mediators of inflammation cause dilation and increased permeability of blood vessels in the nasal cavity. Fluid moves from the blood into tissue, resulting in edema, or swelling. The swollen tissue can restrict air movement, producing a "stopped up" nose. Tissue damage and leakage of fluid into the nasal cavity can also result in a buildup of mucus that blocks the air passageways and eventually results in a runny nose.

5. Cartilage heals slowly after an injury because it has no blood supply; thus cells and nutrients necessary for tissue repair do not easily reach the damaged area. Torn cartilage in the knee causes pain, irritation to the joint, and can lead to arthritic conditions if not removed.

The Integumentary System

FOCUS: The integumentary system consists of the skin, hair, nails, and a variety of glands. The epidermis of the skin provides protection against abrasion, ultraviolet light, and water loss, and produces vitamin D. The dermis provides structural strength and contains blood vessels involved in temperature regulation. The skin is attached to underlying tissue by the hypodermis, which is a major site of fat storage.

CONTENT LEARNING ACTIVITY

Skin

❝*The skin is made up of two major tissue layers, the dermis and the epidermis.*❞

A. Match these terms with the correct statement or definition:

Dermis	Hypodermis
Epidermis	Papillae
Fat	Striae

_____ 1. Loose connective tissue that attaches the skin to underlying bone or muscle; not part of the skin; also called subcutaneous tissue.

_____ 2. Functions as padding and insulation; responsible for some of the structural differences between men and women.

_____ 3. Dense connective tissue; responsible for most of the structural strength of the skin; the deep layer of the skin.

_____ 4. Visible lines produced by overstretching of the dermis.

_____ 5. Projections that contain blood vessels and produce fingerprints.

_____ 6. Stratified squamous epithelium which resists abrasion and forms a permeability barrier.

B. Match these terms with the correct statement or definition:

Callus
Corn
Keratin

Lipid
Stratum basale
Stratum corneum

_____ 1. Strata of the epidermis that produces new cells by mitosis.

_____ 2. Strata of the epidermis that contains dead, squamous cells.

_____ 3. Substance responsible for the structural strength of the stratum corneum.

_____ 4. Substance that helps prevent fluid loss through the skin.

_____ 5. Thickened area of stratum corneum produced in response to friction.

_____ 6. Stratum corneum that thickens to form a cone-shaped structure over a bony prominence.

Skin Color

66 *Skin color is determined primarily by pigments in the skin, and by blood circulation through the skin.* 99

Using the terms provided, complete the following statements:

Albinism
Birthmarks
Blue
Cyanosis

Melanin
Melanocytes
Red
Tan

 (1) is a brown-to-black pigment responsible for most skin color. It is produced by (2) in the stratum basale, and is distributed to other epidermal cells. Melanin production is primarily determined by genetic factors and exposure to light. A mutation that prevents the manufacture of melanin is (3) . Increased melanin production in response to ultraviolet light results in a (4) . Blood flowing through the skin produces a (5) color. When blood flow increases, for example during blushing, this color intensifies. A decrease in the blood oxygen content of blood produces a (6) color called (7) . Congenital disorders of blood vessels in the dermis produce (8) .

1. _____

2. _____

3. _____

4. _____

5. _____

6. _____

7. _____

8. _____

Hair

❝_Hair consists of columns of dead epithelial cells filled with keratin._**❞**

Match these terms with the correct statement or definition:

Arrector pili   Hair root
Hair bulb   Hair shaft
Hair follicle

_____ 1. Portion of hair protruding above the surface of the skin.

_____ 2. An extension of the epidermis into the dermis; holds the hair in place.

_____ 3. Source of new epithelial cells following a burn that damages the surface epithelium.

_____ 4. Site of production of the hair.

_____ 5. Smooth muscles cells that cause hair to "stand on end" and also produce "goose bumps."

☞ Hair is produced in cycles that involve a growth stage alternating with a resting stage.

Glands

❝_The major glands of the skin are the sebaceous glands and the sweat glands._**❞**

Match these terms with the correct statement or definition:

Apocrine sweat gland   Sebaceous gland
Merocrine sweat gland   Sebum

_____ 1. Oily, white substance rich in lipids; prevents drying and protects against some bacteria.

_____ 2. Gland that produces sebum; connected by a duct to the hair follicle.

_____ 3. Sweat gland that produces a watery secretion (sweat); opens onto the surface of the skin.

_____ 4. Sweat gland that produces a thick, organic secretion that is broken down by bacteria to produce body odor; opens into the hair follicle.

☞ Detection of increased sweating is used in lie detector (polygraph) tests because sweat gland activity usually increases when a person tells a lie.

Integumentary System Diagram

B. Match these terms with the correct part labeled in Figure 5-1:

Apocrine sweat gland
Arrector pili
Dermis
Epidermis
Hair bulb
Hair follicle
Hair root

Hair shaft
Hypodermis
Merocrine sweat gland
Sebaceous gland
Stratum basale
Stratum corneum

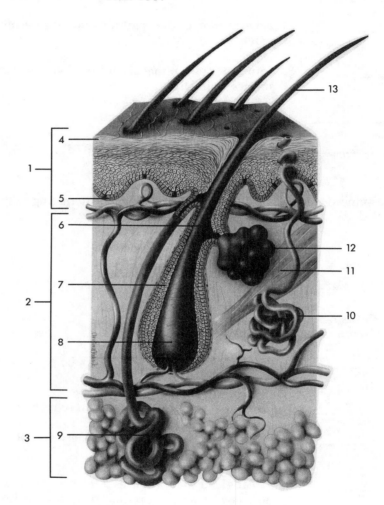

1. _____

2. _____

3. _____

4. _____

5. _____

6. _____

7. _____

8. _____

9. _____

10. _____

11. _____

12. _____

13. _____

Nails

❝*The nail is a thin plate, consisting of dead stratum corneum cells that contain***❞** *a very hard type of the protein keratin.*

A. **M**atch these terms with the correct statement or definition:

Cuticle   Nail matrix
Lunula   Nail root
Nail body

_____ 1. Visible part of the nail.

_____ 2. Stratum corneum that grows onto the nail body.

_____ 3. Produces the nail.

_____ 4. Whitish, crescent-shaped area at the base of a nail; part of the nail matrix.

 Unlike hair, nails grow continuously and do not have a resting stage.

B. **M**atch these terms with the correct part labeled in Figure 5-2:

Cuticle
Lunula
Nail body
Nail root

1. _____

2. _____

3. _____

4. _____

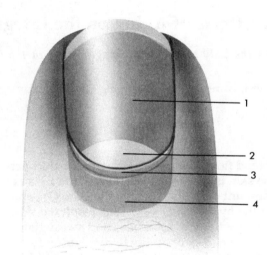

Functions of the Integumentary System

❝*The integumentary system has many functions in the body.***❞**

Match these terms with the
correct statement or definition:

Protection
Sensation

Temperature regulation
Vitamin D production

_____ 1. Accomplished by the skin as a physical barrier that prevents water loss and the entry of microorganisms.

_____ 2. Absorption of ultraviolet light by melanin.

_____ 3. Resists abrasion by sloughing cells from the epidermis.

_____ 4. Carried out by producing sweat and dilating or constricting blood vessels in the dermis.

_____ 5. Begins when an inactive form of the molecule is produced in skin exposed to ultraviolet light.

_____ 6. Detection of pain, heat, cold, and pressure.

The Effects of Aging on the Integumentary System

❝*As the body ages, many changes occur in the integumentary system.***❞**

Using the terms provided, complete the following statements:

Decreases
Increases

As the body ages, blood flow to the skin _(1)_, and the thickness of the skin _(2)_. The skin is more easily damaged and repairs more slowly. The amount of elastic fibers in the skin _(3)_, and the amount of fat in the hypodermis _(4)_. As a result the skin wrinkles and sags. The activity of sebaceous and sweat glands _(5)_, resulting in dry skin and poor ability to regulate body temperature. The number of functioning melanocytes _(6)_, but in some localized areas, especially the hands and face, melanocytes increase in number to produce age spots. White or gray hair also results because melanin production _(7)_.

1. _____

2. _____

3. _____

4. _____

5. _____

6. _____

7. _____

Understanding Burns

"Burns are classified according to the depth of the burn."

Match these terms with the correct statement or definition:

First degree burn
Second degree burn
Third degree burn

Full thickness burn
Partial thickness burn

_____ 1. Some portion of the stratum basale remains viable and regeneration of the epidermis occurs from within the burn area as well as from the edges of the burn; includes first and second degree burns.

_____ 2. Involves only the epidermis; red and painful.

_____ 3. Damages the epidermis and dermis; symptoms include redness, pain, edema, and blisters.

_____ 4. Epidermis and dermis are completely destroyed, and recovery occurs from the edges of the burn; also called third degree burn.

Diseases of the Skin

"Skin cancer is the most common type of cancer."

Match these terms with the correct statement or definition:

Acne
Cold sores
Decubitus ulcer
Herpes zoster (shingles)
Malignant melanoma

Most common cancer
Psoriasis
Ringworm
Rubeola
Rubella

_____ 1. Disorder of the hair follicle and sebaceous gland; produces whiteheads, blackheads, and pimples.

_____ 2. Caused by lack of blood flow to the skin, resulting in skin death followed by infection; produces sores over bony projections.

_____ 3. Type of measles most likely to be damaging to the fetus.

_____ 4. Following recovery from chickenpox, lesions form in the skin along the path of a nerve; results when the dormant chickenpox virus becomes active.

_____ 5. Lesions around the skin of the mouth and in the mucous membranes of the mouth; caused by herpes simplex I virus.

_____ 6. Disorder that produces large, silvery scales that bleed when removed; cause unknown.

_____ 7. Skin cancer that involves the deep cells of the epidermis; produces an open ulcer or nodular tumor.

_____ 8. Rare form of skin cancer that arises from melanocytes, usually in a preexisting mole; unless diagnosed and treated early this cancer is often fatal.

1. Name the two layers of the skin and state what kind of tissue makes up each layer.

2. Name the deepest and most superficial layers of the epidermis and describe the type of cells found in each layer.

3. State the two factors primarily responsible for skin color.

4. Name the two stages in the hair growth cycle.

5. List the three types of glands found in the skin.

6. List four protective functions of the skin.

7. State two ways the integumentary system functions to regulate body temperature.

Place the letter corresponding to the correct answer in the space provided.

_____ 1. The hypodermis
   a. connects the dermis to underlying bone and muscle.
   b. is the layer of skin where hair is produced.
   c. is the layer of skin where nails are produced.
   d. connects the dermis and the epidermis.

_____ 2. The part of the skin where cells divide by mitosis in order to replace cells lost from the outermost surface of the skin?
   a. hypodermis
   b. dermis
   c. stratum basale
   d. stratum corneum

_____ 3. The dermis
   a. is loose connective tissue.
   b. contains approximately half of the body's fat deposits.
   c. has papillae which are responsible for fingerprints.
   d. does not contain blood vessels.

_____ 4. In what area of the body would you expect to find an especially thick stratum corneum?
   a. back of the hand
   b. heel of the foot
   c. abdomen
   d. over the shin

_____ 5. The function of keratin in the skin is
   a. lubrication of the skin.
   b. to reduce water loss.
   c. to provide protection from ultraviolet light.
   d. to provide structural strength.

_____ 6. Concerning skin color, which of the following statements is NOT correctly matched?
   a. no skin pigmentation (albinism) - genetic disorder
   b. skin tans - increased melanin production
   c. skin appears blue (cyanosis) - oxygenated blood
   d. negroes darker than caucasians - more melanin in negroes

_____ 7. Hair
   a. slowly, but continually grows.
   b. grows from the tip of the hair shaft.
   c. consists of columns of dead epithelial cells filled with keratin.
   d. all of the above

_____ 8. A hair follicle
   a. is an extension of the epidermis into the dermis.
   b. receives a duct from a sebaceous gland.
   c. receives a duct from an apocrine sweat gland.
   d. all of the above

_____ 9. Smooth muscles that produce "goose bumps" when they contract are the
   a. papillae.
   b. cuticle.
   c. hypodermal muscles.
   d. arrector pili.

_____ 10. Sebum
   a. lubricates hair and skin, which prevents drying.
   b. is produced by sweat glands.
   c. consists of dead cells from hair follicles.
   d. is responsible for body odor.

_____ 11. To increase heat loss form the body, one would expect _____ of dermal blood vessels and _____ sweating.
a. dilation, decreased
b. dilation, increased
c. constriction, decreased
d. constriction, increased

_____ 12. While building the patio deck to his house, an anatomy and physiology instructor hit his finger with a hammer. He responded by saying, "Gee, I hope I didn't irreversibly damage the _____, because if I did, my fingernail will never grow back."
a. cuticle
b. nail body
c. nail matrix
d. nail root

_____ 13. Skin aids in maintaining the calcium and phosphate levels of body by participating in the production of
a. sebum.
b. keratin.
c. vitamin A.
d. vitamin D.

_____ 14. On a sunny spring day a student decided to initiate her annual tanning ritual. However, she fell asleep while sunbathing. After awakening she noticed that the skin on her back was burned. She experienced redness, blisters, edema, and pain. The burn was nearly healed about 10 days later. The burn was best classified as a
a. first degree burn.
b. second degree burn.
c. third degree burn.

FINAL CHALLENGES

Use a separate sheet of paper to complete this section.

1. The skin of infants is more easily penetrated and injured by abrasion than that of adults. What part of the epidermis is probably much thinner in infants than in adults?

2. The rate of water loss from the skin of the hand was measured. Following the measurement the hand was soaked in alcohol for 15 minutes. Alcohol has the ability to dissolve lipids. After all the alcohol was removed from the hand, the rate of water loss was again measured. Compared to the rate of water loss before soaking the hand in alcohol, what difference, if any, would you expect in the rate of water loss after soaking the hand in alcohol.

3. How would a person born without any sweat glands be different from a person with sweat glands?

4. You may have noticed that on very cold winter days people's noses and ears turn red. Can you explain why this happens?

5. Given what is known about the cause of body odor, propose some ways to prevent the condition.

6. Dandy Chef has been burned on the arm. The doctor, using a forceps, pulls on a hair within the area that was burned. The hair easily pulls out. What degree of burn did the patient have and how do you know?

ANSWERS TO CHAPTER 5

Skin
   A. 1. Hypodermis; 2. Fat; 3. Dermis; 4. Striae;
       5. Papillae; 6. Epidermis
   B. 1. Stratum basale; 2. Stratum corneum;
       3. Keratin; 4. Lipid; 5. Callus; 6. Corn

Skin Color
   1. Melanin; 2. Melanocytes; 3. Albinism; 4. Tan;
   5. Red; 6. Blue; 7. Cyanosis; 8. Birthmarks

Hair
   1. Hair shaft; 2. Hair follicle; 3. Hair follicle; 4. Hair
   bulb; 5. Arrector pili

Glands
   1. Sebum; 2. Sebaceous gland; 3. Merocrine sweat
   gland; 4. Apocrine sweat gland

Integumentary System Diagram
   1. Epidermis; 2. Dermis; 3. Hypodermis; 4. Stratum
   corneum; 5. Stratum basale; 6. Hair root; 7. Hair
   follicle; 8. Hair bulb; 9. Apocrine sweat gland;
   10. Merocrine sweat gland; 11. Arrector pili;
   12. Sebaceous gland; 13. Hair shaft

Nails
   A. 1. Nail body; 2. Cuticle; 3. Nail matrix;
       4. Lunula
   B. 1. Nail body; 2. Lunula; 3. Cuticle; 4. Nail root

Functions of the Integumentary System
   1. Protection; 2. Protection; 3. Protection;
   4. Temperature regulation; 5. Vitamin D production;
   6. Sensation

The Effects of Aging on the Integumentary System
   1. Decreases; 2. Decreases; 3. Decreases;
   4. Decreases; 5. Decreases; 6. Decreases;
   7. Decreases

Understanding Burns
   1. Partial thickness burn; 2. First degree burn;
   3. Second degree burn; 4. Full thickness burn

Diseases of the Skin
   1. Acne; 2. Decubitus ulcer; 3. Rubella; 4. Herpes
   zoster (shingles); 5. Cold sores; 6. Psoriasis; 7. Most
   common cancer; 8. Malignant melanoma

1. Deepest layer is the dermis, which consists of dense
    connective tissue; superficial layer is the epidermis,
    which consists of stratified squamous epithelium
2. Deepest layer is the stratum basale, which consists
    of columnar cells that divide to produce new cells;
    most superficial layer is the stratum corneum, which
    consists of dead, squamous cells filled with keratin
3. Pigments (melanin) and blood

4. Growth stage and resting stage
5. Sebaceous glands, merocrine sweat glands, and
    apocrine sweat glands
6. Protects against abrasion, ultraviolet light, water loss,
    entry of microorganisms
7. Increasing or decreasing blood flow through the skin;
    increased sweat production

1. A.   The hypodermis connects the dermis to
    underlying bone and muscle. It is not part of the
    skin. Hair and nails are derived from the epidermis.

2. C.   Epidermal cells are produced in the stratum
    basale and move upward into the stratum corneum.
    During movement the cells change shape, become
    filled with keratin, and die.

3. C.   The dermis has projections, called papillae,
    which extend into the epidermis. The papillae
    contain blood vessels that supply the epidermis. The
    papillae also form fingerprints. The dermis is dense

connective tissue. The hypodermis contains
approximately half the body's fat.

4. B.   The heel of the foot is subjected to friction and
    has a thick stratum corneum. A callus or corn is
    possible.

5. D.   Keratin provides structural strength, lipids
    prevent water loss, melanin protects against
    ultraviolet light, and sebum lubricates the skin.

6. C.   Cyanosis, a bluish skin color, is caused by
    deoxygenated blood.

7. C. Hair is columns of dead epithelial cells filled with keratin. Hair is produced in the hair bulb (not the tip of the hair shaft). Hair grows in cycles: a growth stage followed by a resting stage.

8. D. Hair follicles are extensions of the epidermis into the dermis. Both sebaceous and apocrine sweat glands empty into hair follicles.

9. D. Arrector pili are smooth muscle that produce "goose bumps" and make hair "stand on end."

10. A. Sebum lubricates hair and skin. It is a lipid produced by sebaceous glands. Merocrine sweat glands produce a watery secretion (sweat) and secretions of apocrine sweat glands are involved in producing body odor.

11. B. Dilation of dermal blood vessels increases blood flow through the skin. Consequently heat is carried from deeper in the body to the surface at which it is lost. Increased sweating results in heat loss as the sweat evaporates.

12. C. The nail grows from the nail matrix. Specifically, the nail matrix forms the nail root, which is pushed distally to become the nail body.

13. D. Vitamin D stimulates calcium and phosphate uptake in the small intestine. Exposure of the skin to ultraviolet light results in the formation of an inactive form of Vitamin D. The inactive form is modified by the liver and kidney into active vitamin D.

14. B. Second degree burns are characterized by redness, blisters, edema, and pain.

 FINAL CHALLENGES

1. In adults, the stratum corneum consists of many layers of squamous, dead cells filled with keratin. These cells provide protection against abrasion. In infants, there are fewer layers of cells in the stratum corneum.

2. Alcohol dissolves lipids, and it removes the lipids from the skin, especially in the stratum corneum. Because these lipids normally prevent water loss, after soaking the hand in alcohol, the rate of water loss can be expected to increase.

3. A lack of sweat glands results in an inability to produce sweat. Consequently there is a decreased ability to regulate body temperature in warm environments.

4. The red color indicates increased blood flow to the tissues. The blood is warm and protects the tissue from cold damage.

5. Because body odor results from the breakdown of the organic secretions of apocrine sweat glands, one possibility is to remove the secretions by washing. Another method is to kill the bacteria. The aluminum salts in some antiperspirants do this. Antiperspirants also reduce the watery secretions of merocrine sweat glands, but these secretions are not the cause of body odor. Deodorants mask body odor with another scent but do not prevent it.

6. The hair pulling out easily indicates that the hair follicle has been destroyed. Because the hair follicle extends deep into the dermis, this would indicate the epidermis and dermis have been destroyed. It is probably a third degree burn.

The Skeletal System — Bones and Joints

FOCUS: The skeletal system consists of bone, cartilage, and ligaments. Bone has several important functions, including support, protection, mineral and fat storage, and blood cell production. Osteocytes, or bone cells, are surrounded by an extracellular matrix containing collagen, which lends flexible strength, and minerals, which give bone weight-bearing strength. The two major types of bone are compact bone and cancellous bone. Bone formation, growth, remodeling, and repair are dynamic processes carried out by osteoblasts and osteoclasts. The skeletal system consists of the axial skeleton (skull, vertebral column, and rib cage) and the appendicular skeleton (limbs and their girdles). The skull surrounds and protects the brain; the vertebral column supports the head and trunk and protects the spinal cord; and the rib cage protects the heart and lungs. The pectoral girdle attaches the upper limbs to the trunk. The pelvic girdle attaches the lower limbs to the trunk. Three major classes of joints are fibrous, cartilaginous, and synovial. Synovial joints are the most freely moveable joints, and are classified according to the type of movement that occurs.

CONTENT LEARNING ACTIVITY

Functions of the Skeletal System

66 *The skeletal system provides support and protection, allows body movements, stores minerals* 99
and fats, and is the site of blood cell production.

Match these terms with the correct statement or definition:

Bone
Cartilage

Ligaments
Tendons

_____ 1. Provides strong rigid support, storage, and protection for the body.

_____ 2. Provides firm, yet flexible support within nose and ear.

_____ 3. Attaches skeletal muscle to bone.

General Features of Bone

"There are four types of bone, described by their shape as long, short, flat, and irregular."

Using the terms provided, complete the following statements:

Articular cartilage   Periosteum
Diaphysis   Red marrow
Epiphysis   Yellow marrow
Medullary cavity

Each long bone consists of a shaft, called the _(1)_, and a(n) _(2)_ at each end of the bone. The epiphyses in joints are covered by _(3)_, and the remaining bone surfaces are covered by a connective tissue membrane, the _(4)_, which contains blood vessels and nerves. The large cavity in the diaphysis is called the _(5)_, and there is a network of smaller spaces in the epiphyses. These spaces are filled with _(6)_, which is mostly fat, or _(7)_, which consists of blood-forming cells.

1. _____
2. _____
3. _____
4. _____
5. _____
6. _____
7. _____

Bone Histology

"Bone tissue consists of cells surrounded by a matrix that contains collagen and minerals."

A. Match these terms with the correct statement or definition:

Compact bone   Osteocytes
Lacunae   Spongy Bone

_____ 1. Bone cells.

_____ 2. Spaces in bone matrix that contain osteocytes.

_____ 3. Interconnecting plates of bone; forms the epiphyses of long bones and the interior of other bones.

_____ 4. Solid matrix of bone containing osteocytes; forms the diaphysis and outer covering of bones.

B. Using the terms provided, complete the following statements:

Canaliculi   Matrix
Haversian canals   Nutrients

Compact bone consists of osteocytes located between thin sheets of _(1)_. Osteocytes are connected to each other by cell processes that extend through small canals called _(2)_. Osteocytes receive nutrients from blood vessels that run parallel to the long axis of the bone through _(3)_. Thus, _(4)_ leave the vessels of the haversian canals and diffuse to the osteocytes through the canaliculi; waste products diffuse in the opposite direction.

1. _____
2. _____
3. _____
4. _____

Bone Formation and Growth

66_Before birth, the skeleton begins as either fibrous membranes or cartilage models._**99**

Match these terms with the
correct statement or definition:

Epiphyseal plate   Osteocytes
Fontanels (soft spots)   Periosteum
Osteoblasts

_____ 1. Bone forming cells.

_____ 2. Bone cells that are completely surrounded by bone matrix.

_____ 3. Portions of the membrane around the brain that are not ossified at
                              birth.

_____ 4. Increase in bone diameter occurs when bone matrix is added beneath
                              this membrane.

_____ 5. Location for increase in bone length; a layer of cartilage between the
                              diaphysis and epiphysis.

Bone Remodeling and Repair

66_Bone growth, remodeling, and repair all involve deposition of new bone matrix by osteoblasts._**99**

Using the terms provided, complete the following statements:

Blood   Medullary
Calcium ion   Osteoblasts
Callus   Osteoclasts
Clot

Bone remodeling involves the removal of old bone by cells
called _(1)_, and the deposition of new bone by osteoblasts. Bone
remodeling occurs during bone growth or _(2)_ regulation. When
a bone becomes larger, the size of the _(3)_ cavity increases,
making the bone lighter. When _(4)_ calcium levels decrease
below normal, osteoclasts break down bone and release calcium
into the blood. When a bone is broken, blood vessels bleed, and
a _(5)_ is formed. Then a fibrous network with cartilage called a
_(6)_ is produced, which helps to hold the bone fragments
together. _(7)_ then enter the callus and begin forming bone.

1. _____

2. _____

3. _____

4. _____

5. _____

6. _____

7. _____

Disorders of Bone

Many types of disorders can occur in bone development, growth, or maintenance.

A. **M**atch these terms with the correct statement or definition:

Dwarfism   Osteogenesis imperfecta
Giantism   Rickets

_____ 1. Abnormally increased height, usually caused by overproduction of growth hormone.

_____ 2. Abnormal shortness, often a result of improper growth in the epiphyseal plates of long bones.

_____ 3. Group of genetic disorders that produce bones that are brittle and easily fractured because of insufficient collagen formation.

_____ 4. Growth retardation resulting from nutritional deficiencies of calcium or vitamin D.

B. **M**atch these terms with the correct definition or description:

Osteomalacia   Osteoporosis
Osteomyelitis

_____ 1. Bone inflammation, often resulting from bacterial infection.

_____ 2. Softening of bones resulting from calcium depletion, e. g., during pregnancy.

_____ 3. Porous bone that results when bone is broken down faster than it is produced.

C. **M**atch these fractures with the correct definition or description:

Closed   Linear
Comminuted   Oblique
Complete   Open
Incomplete   Transverse

_____ 1. Fracture in which the bone protrudes through the skin.

_____ 2. Fracture in which the bone fragments are not completely separated.

_____ 3. Fracture in which the bone breaks into more than two fragments.

_____ 4. Fracture that is parallel to the long axis of the bone.

_____ 5. Fracture that is at an angle other than a right angle to the long axis of the bone.

Skull

"The bones of the skull are divided into brain case bones, facial bones, and the auditory ossicles."

A. Match these terms with the correct statement or definition:

Auditory ossicles	Hyoid bone
Brain case	Sutures
Facial bones	

_____ 1. Part of the skull that protects the brain.

_____ 2. Bones that form the structure of the face.

_____ 3. Bones that transmit sound waves to the inner ear.

_____ 4. Small U-shaped bone in the neck; the attachment site for several neck and tongue muscles.

_____ 5. Fibrous connections between bones of the skull.

B. Match these terms with the correct statement or definition:

Hard palate	Nasal septum
Mastoid air cells	Orbit
Nasal conchae	Paranasal sinuses

_____ 1. Eye socket.

_____ 2. Divides the nasal cavity into right and left halves.

_____ 3. Three bony shelves on the lateral wall of the nasal cavity; function to increase surface area.

_____ 4. Four large cavities that open into the nasal cavity.

_____ 5. Small cavities inside the mastoid process of the temporal bone.

_____ 6. Floor of the nasal cavity; formed from the maxillae and palatine bones.

C. **Match these openings or depressions with the correct description:**

Carotid canal   Mandibular fossa
Cranial fossae   Nasolacrimal canal
External auditory meatus   Optic foramen
Foramen magnum   Sella turcica
Jugular foramen

_____ 1. Canal that enables sound waves to reach the eardrum.

_____ 2. Opening through which the optic nerve passes into the brain case.

_____ 3. Passageway that contains a duct which carries excess tears from the eye to the nasal cavity.

_____ 4. Opening through which the spinal cord connects to the brain.

_____ 5. In addition to the foramen magnum, the passageways for entry of major blood vessels into the skull.

_____ 6. Opening in the base of the skull for blood leaving the brain.

_____ 7. Depression where the mandible articulates with the temporal bone.

_____ 8. Depressions that hold the brain; located in the floor of the brain case.

_____ 9. Depression in the sphenoid bone that resembles a saddle; holds the pituitary gland.

D. **Match these bone parts with structures to which they contribute:**

Palatine   Temporal
Maxilla   Vomer
Ethmoid   Zygomatic

_____ 1. Parts of these two bones form the hard palate.

_____

_____ 2. Parts of these two bones form the nasal septum.

_____

_____ 3. Parts of these two bones form the zygomatic arch.

_____

☞ The axial skeleton includes the bones of the head and trunk, and the appendicular skeleton consists of the limb bones and their girdles.

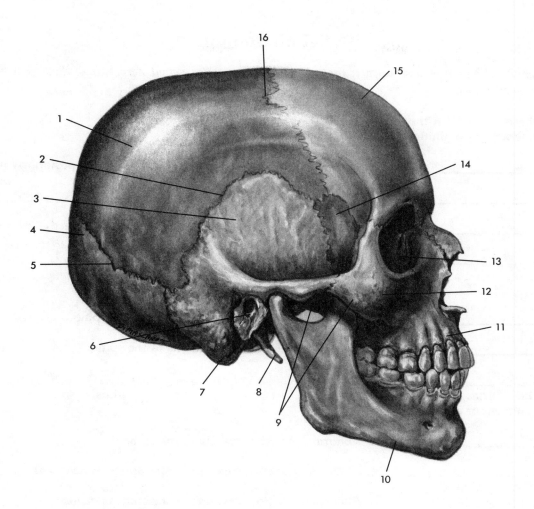

E. **Match these terms with the correct parts labeled in Figure 6-1:**

Coronal suture
External auditory meatus
Frontal bone
Lambdoid suture
Mandible
Mastoid process
Maxilla
Nasolacrimal canal

Occipital bone
Parietal bone
Sphenoid bone
Squamous suture
Styloid process
Temporal bone
Zygomatic arch
Zygomatic bone

1. _____

2. _____

3. _____

4. _____

5. _____

6. _____

7. _____

8. _____

9. _____

10. _____

11. _____

12. _____

13. _____

14. _____

15. _____

16. _____

Vertebral Column

66 *The vertebral column provides support, protection, an exit site for spinal nerves, and allows movement* 99
of the head and trunk.

A. Match these terms with the
correct statement or definition:

Cervical   Sacral
Lumbar   Thoracic

_____ 1. Two sections of the vertebral column that curve posteriorly in adults.

_____

_____ 2. Two sections of the vertebral column with secondary curvatures.

_____

_____ 3. There are seven of these vertebrae in the spine.

_____ 4. There are twelve of these vertebrae in the spine.

_____ 5. There are five of these vertebrae in the spine.

B. Match these terms with the
correct statement or definition:

Kyphosis   Scoliosis
Lordosis

_____ 1. Abnormal lateral curvature of the spine.

_____ 2. Abnormal posterior curvature of the spine (hunchback).

_____ 3. Abnormal anterior curvature of the spine (swayback).

Understanding the Vertebral Column

66 *The vertebral column illustrates the relationship between structure and function.* 99

A. Match these terms with the
correct statement or definition:

Anulus fibrosus   Intervertebral disk
Body   Nucleus pulposus
Herniated disk

_____ 1. Bony cylinder that is the weight-bearing portion of the vertebra.

_____ 2. Fibrocartilage structure located between the bodies of adjacent vertebrae.

_____ 3. External fibrous ring of the intervertebral disk.

_____ 4. Internal gelatinous material of the intervertebral disk.

_____ 5. Condition in which the anulus fibrosus ruptures, with a partial or complete release of the nucleus pulposus.

B. Match these terms with the correct statement or definition:

Intervertebral foramen   Vertebral canal
Spina bifida   Vertebral foramen
Vertebral arch

_____ 1. Opening in a vertebra, through which the spinal cord passes.

_____ 2. Passageway formed by the vertebral foramen in adjacent vertebrae.

_____ 3. Part of a vertebra that surrounds the vertebral foramen.

_____ 4. Condition in which vertebral arches are absent or incompletely formed

_____ 5. An opening formed by notches in the vertebral arches of adjacent vertebrae; exit site for spinal nerves.

C. Match these terms with the correct statement or definition:

Spinous processes   Transverse processes
Superior and inferior
   articular processes

_____ 1. Processes that allow vertebrae to join with other vertebrae above and below them.

_____ 2. Processes that extend laterally from each side of the vertebral arch and serve as points of muscle attachment.

_____ 3. Processes that extend posteriorly from the vertebral arch and serve as points of muscle attachment.

Thoracic Cage

"_The thoracic cage, or rib cage, protects the vital organs within the thorax and prevents the collapse_**"** _of the thorax during respiration._

Match these terms with the correct statement or definition:

Costal cartilage   Sternum
False ribs   True ribs
Floating ribs   Xiphoid process

_____ 1. Superior seven pairs of ribs that attach directly to the sternum.

_____ 2. Structure by which a rib is attached to the sternum.

_____ 3. Inferior five pairs of ribs.

_____ 4. Eleventh and twelfth ribs; have no attachment to the sternum.

_____ 5. The breastbone.

_____ 6. Most inferior part of the sternum.

 All of the ribs attach to the thoracic vertebrae.

71

Pectoral Girdle

*66**The girdles attach the limbs to the axial skeleton.**99*

Match these terms with the
correct statement or definition:

Clavicle
Scapula

_____ 1. The shoulder blade; attachment site for muscles that move the arm.

_____ 2. The collar bone; attaches the scapula to the sternum.

Upper Limb

*66**The upper limb consists of the bones of the arm, forearm, wrist, and hand.**99*

Match these terms with the
correct statement or definition:

Carpals   Phalanges
Humerus   Radius
Metacarpals   Ulna

_____ 1. Bone that articulates with the scapula and forearm bones.

_____ 2. Bone on the medial (little finger) side of the forearm.

_____ 3. Eight bones found in the wrist.

_____ 4. Five bones that form the palm of the hand.

_____ 5. Three small bones found in each finger (two in the thumb).

Pelvic Girdle

*66**The pelvic girdle is a ring of bones formed by the sacrum and two coxae.**99*

Using the terms provided, complete the following statements:

Acetabulum   Male
Anterior superior iliac spine   Obturator foramen
Coxa   Pelvic brim
Female   Pubic symphysis
Iliac crest   Sacroiliac joint

Each _(1)_ is formed by the fusion of the ilium, ischium, and
pubis. The superior margin of the ilium is called the _(2)_, and
an _(3)_ is located at each ilium's anterior end. The coxae join
each other at the _(4)_, and join the sacrum at the _(5)_. The
_(6)_ is the socket of the hip joint, and the _(7)_ is the large hole
in the coxa. The opening of the pelvic cavity, the _(8)_, is larger
and more oval in the pelvis of the _(9)_.

1. _____

2. _____

3. _____

4. _____

5. _____

6. _____

7. _____

8. _____

9. _____

Location of the Major Bones of the Skeletal System

Match these bones with
the correct parts labeled
in Figure 6-2:

Carpals
Clavicle
Coxa
Femur
Fibula
Humerus
Metacarpals
Metatarsals
Patella
Phalanges
Radius
Scapula
Tarsals
Tibia
Ulna

1. _____

2. _____

3. _____

4. _____

5. _____

6. _____

7. _____

8. _____

9. _____

10. _____

11. _____

12. _____

13. _____

14. _____

15. _____

16. _____

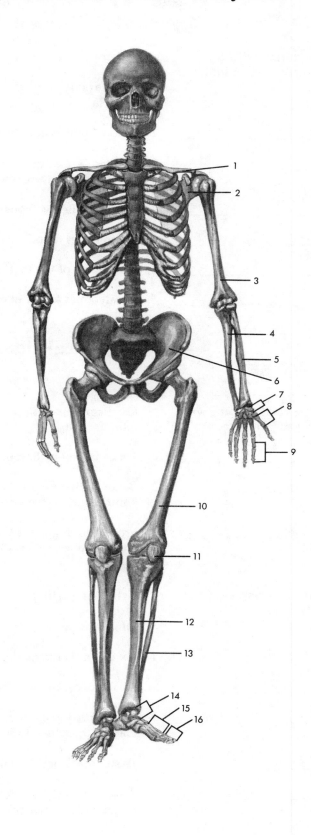

Lower Limb

" *The lower limb consists of the bones of the thigh, leg, ankle, and foot.* **"**

Match these terms with the
correct statement or definition:

Femur   Phalanges
Fibula   Tarsals
Metatarsals   Tibia
Patella

_____ 1. The only bone in the thigh.

_____ 2. Located within the major tendon of the thigh muscles; enables the tendon to turn the corner over the knee.

_____ 3. Larger of the two bones in the leg; the shin bone.

_____ 4. Seven bones found in the ankle.

_____ 5. Five bones of the foot; the distal ends make up the ball of the foot.

_____ 6. Small bones found in the toes.

Articulations

" *An articulation, or joint, is a place where two bones come together.* **"**

A. Match the class of joint with
the correct definition:

Cartilaginous joint   Synovial joint
Fibrous joint

_____ 1. Bones united by fibrous tissue; little or no movement, e.g., sutures.

_____ 2. Bones united by cartilage; only slight movement can occur at these joints, e.g., intervetebral disks and the pubic symphysis.

_____ 3. Freely moving joints that contain fluid in a cavity surrounding the ends of bones, e.g., the knee and elbow.

B. Match these terms with the
correct statement or definition:

Articular cartilage   Joint cavity
Bursa   Synovial membrane
Joint capsule

_____ 1. Cartilage that provides a smooth surface where bones meet.

_____ 2. Small space surrounding the ends of articulating bones.

_____ 3. Surrounds the joint cavity and helps hold the bones together; portions may be thickened to form ligaments.

_____ 4. Tissue that lines the joint capsule; produces synovial fluid.

_____ 5. Extension of the synovial membrane that forms a pocket or sac; reduces friction where structures would rub together.

C. Match these terms with the correct
   parts labeled in Figure 6-3:

   Articular cartilage
   Bursa
   Joint capsule
   Joint cavity
   Synovial membrane

   1. _____

   2. _____

   3. _____

   4. _____

   5. _____

Tendon

Types of Synovial Joints

"Synovial joints are classified according to the shape of the adjoining articular surfaces."

Match these terms with
the correct parts labeled
in Figure 6-4:

Ball-and-socket joint
Ellipsoid (condyloid) joint
Hinge joint

Pivot joint
Plane joint
Saddle joint

1. _____   3. _____   5. _____

2. _____   4. _____   6. _____

Types of Movement

The types of movement occurring at a given joint are related to the structure of that joint.

Match these terms with the correct parts labeled in Figure 6-5:

Abduction
Adduction
Circumduction
Extension

Flexion
Pronation
Supination

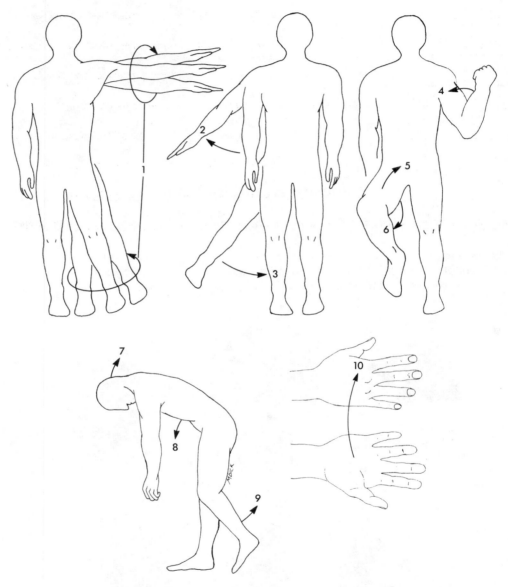

1. _____

2. _____

3. _____

4. _____

5. _____

6. _____

7 _____

8. _____

9. _____

10. _____

Joint Disorders

"Joint disorders can occur from inflammation or trauma."

Match these terms with the correct statement or definition:

Bursitis
Bunion
Degenerative Joint Disease
   (Osteoarthritis)
Dislocation

Gout
Rheumatoid arthritis
Separation
Sprain

_____ 1. Autoimmune disease; articular cartilage of a joint is destroyed.

_____ 2. Articular cartilage loss; result of "wear and tear" of advancing age.

_____ 3. Inflammation of toe, foot, and leg joints caused by increased uric acid.

_____ 4. Inflammation of a bursa caused by injury or infection.

_____ 5. Bursitis that develops over the joint at the base of the great toe.

_____ 6. Tearing or pulling of the ligaments around a joint.

_____ 7. Condition in which bones remain apart after an injury to a joint.

_____ 8. Condition in which the head of one bone is pulled out of the socket of another bone.

1. List three types of bone cells, depending on their function.

2. List two types of bone, according to their internal structure.

3. Name the five types of vertebrae, and give the number of each found in the vertebral column.

4. Name three types of ribs according to their attachment, and give the number of each type.

5. Give the number of carpals, metacarpals, and phalanges in the upper limb, and give the number of tarsals, metatarsals, and phalanges in the lower limb.

6. List the three major classes of joints.

7. Name the six types of synovial joints and give an example of each.

8. List eight types of movements at joints.

MASTERY LEARNING ACTIVITY

Place the letter corresponding to the correct answer in the space provided.

_____ 1. Which of the following is a function of bone?
- a. internal support and protection
- b. provide attachment for muscles
- c. mineral storage
- d. blood cell formation
- e. all of the above

_____ 2. Concerning bone matrix,
- a. collagen provides flexible strength.
- b. calcium and phosphorus provide weight-bearing strength.
- c. osteocytes are surrounded by matrix.
- d. tiny canals called canaliculi pass through the matrix of compact bone.
- e. all of the above

_____ 3. A break in the shaft of a bone would be a break in the
- a. epiphysis.
- b. articular cartilage.
- c. diaphysis.
- d. fontanel.

_____ 4. Which of the following connective tissue structures covers the surface of mature bones?
- a. callus
- b. periosteum
- c. cartilage
- d. b and c

_____ 5. In compact bone, the osteocytes are connected to each other by cell processes extending through tiny canals called
- a. periosteum.
- b. lacunae.
- c. haversian canals.
- d. canaliculi.

_____ 6. Fontanels are found
- a. at the epiphyseal plate.
- b. primarily in long bones.
- c. in skull bones formed from membranes.
- d. in the diaphysis.
- e. both a and b

_____ 7. The prime function of osteoclasts is to
- a. prevent osteoblasts from forming.
- b. break down bone.
- c. produce calcium salts and collagen fibers.
- d. change spongy bone to cartilage.

_____ 8. In the healing of bone fractures
- a. a blood clot forms around the break.
- b. a callus is formed.
- c. bone is formed in the callus.
- d. the callus may eventually disappear.
- e. all of the above

_____ 9. Bone inflammation that often results from bacterial infection is called
- a. osteogenesis imperfecta.
- b. osteomyelitis.
- c. rickets.
- d. osteoporosis.
- e. osteomalacia

_____ 10. Which of the following is part of the appendicular skeleton?
- a. skull
- b. ribs
- c. clavicle
- d. sternum
- e. vertebra

_____ 11. The perpendicular plate of the ethmoid and the _____ form the nasal septum.
- a. zygomatic arch
- b. nasal bone
- c. nasal conchae
- d. vomer

_____ 12. Which of the following bones does NOT contain a paranasal sinus?
- a. ethmoid
- b. sphenoid
- c. temporal
- d. frontal
- e. maxilla

_____ 13. The passageway that carries tears from the eyes to the nasal cavity is the
a. nasolacrimal canal.
b. optic foramen.
c. paranasal sinuses.
d. foramen magnum.

_____ 14. Which of the following parts of the upper limb is NOT correctly matched with the number of bones in that part?
a. arm: 1
b. forearm: 2
c. wrist: 10
d. palm of hand: 5
e. fingers: 14

_____ 15. Which of the following is characteristic of a synovial joint?
a. articular surfaces covered with cartilage
b. joint capsule
c. synovial membrane
d. synovial fluid
e. all of the above

_____ 16. Once a doorknob is grasped, what movement of the forearm is necessary to unlatch the door (turn in a clockwise direction)?
a. pronation
b. rotation
c. flexion
d. supination
e. extension

_____ 17. An autoimmune disease that destroys the articular cartilage in joints is
a. rheumatoid arthritis.
b. degenerative joint disease.
c. gout.
d. bursitis.
e. osteoarthritis.

☆ ——— FINAL CHALLENGES ——— ☆

Use a separate sheet of paper to complete this section.

1. In what region of the vertebral column is a ruptured disk most likely to develop? Explain.

2. The length of the lower limbs of a 5-year-old were measured, and it was determined that one limb was 2 cm shorter than the other limb. Would you expect the little girl to exhibit kyphosis, scoliosis, or lordosis? Explain.

3. Can you suggest a possible advantage in having the coccyx attached to the sacrum by a flexible cartilage joint?

4. What would be the likely consequences if a fontanel in a baby's skull fused shortly after birth?

5. Seat belts should be fastened as low as possible across the hips. Defend or refute this statement.

ANSWERS TO CHAPTER 6

Functions of the Skeletal System
1. Bone; 2. Cartilage; 3. Tendons

General Features of Bone
1. Diaphysis; 2. Epiphysis; 3. Articular cartilage;
4. Periosteum; 5. Medullary cavity; 6. Yellow marrow;
7. Red marrow

Bone Histology
A. 1. Osteocytes; 2. Lacunae; 3. Spongy bone;
4. Compact bone
B. 1. Matrix; 2. Canaliculi; 3. Haversian canals;
4. Nutrients

Bone Formation and Growth
1. Osteoblasts; 2. Osteocytes; 3. Fontanels (soft
spots); 4. Periosteum; 5. Epiphyseal plate

Bone Remodeling and Repair
1. Osteoclasts; 2. Calcium ion; 3. Medullary;
4. Blood; 5. Clot; 6. Callus; 7. Osteoblasts;

Bone Disorders
A. 1. Giantism; 2. Dwarfism; 3. Osteogenesis
imperfecta; 4. Rickets
B. 1. Osteomyelitis; 2. Osteomalacia;
3. Osteoporosis
C. 1. Open; 2. Incomplete; 3. Comminuted;
4. Linear; 5. Oblique

Skull

A. 1. Brain case; 2. Facial bones; 3. Auditory ossicles;
4. Hyoid bone; 5. Sutures
B. 1. Orbit; 2. Nasal septum; 3. Nasal conchae;
4. Paranasal sinuses; 5. Mastoid air cells; 6. Hard
palate
C. 1. External auditory meatus; 2. Optic foramen;
3. Nasolacrimal canal; 4. Foramen magnum;
5. Carotid canal; 6. Jugular foramen;
7. Mandibular fossa; 8. Cranial fossae; 9. Sella
turcica
D. 1. Maxilla, Palatine; 2. Ethmoid, Vomer;
3. Zygomatic, Temporal
E. 1. Parietal bone; 2. Squamous suture;
3. Temporal bone; 4. Occipital bone; 5. Lambdoid
suture; 6. External auditory meatus; 7. Mastoid
process; 8. Styloid process; 9. Zygomatic arch;
10. Mandible; 11. Maxilla; 12. Zygomatic bone;
13. Nasolacrimal canal; 14. Sphenoid bone;
15. Frontal bone; 16. Coronal suture

Vertebral Column
A. 1. Thoracic, Sacral; 2. Cervical, Lumbar;
3. Cervical; 4. Thoracic; 5. Lumbar
B. 1. Scoliosis; 2. Kyphosis; 3. Lordosis

Unterstanding the Vertebral Column
A. 1. Body; 2. Intervertebral disk; 3. Anulus fibrosus;
4. Nucleus pulposus; 5. Herniated disk
B. 1. Vertebral foramen; 2. Vertebral canal;
3. Vertebral arch; 4. Spina bifida;
5. Intervertebral foramen
C. 1. Superior and inferior articular processes;
2. Transverse processes; 3. Spinous processes

Thoracic Cage
1. True ribs; 2. Costal cartilage; 3. False ribs;
4. Floating ribs; 5. Sternum; 6. Xiphoid process

Pectoral Girdle
1. Scapula; 2. Clavicle

Upper Limb
1. Humerus; 2. Ulna; 3. Carpals; 4. Metacarpals;
5. Phalanges

Pelvic Girdle
1. Coxa; 2. Iliac crest; 3. Anterior superior iliac spine;
4. Pubic symphysis; 5. Sacroiliac joint;
6. Acetabulum; 7. Obturator foramen; 8. Pelvic brim;
9. Female

Location of the Major Bones of the Skeletal System
1. Clavicle; 2. Scapula; 3. Humerus; 4. Ulna;
5. Radius; 6. Coxa; 7. Carpals; 8. Metacarpals;
9. Phalanges; 10. Femur; 11. Patella; 12. Tibia;
13. Fibula; 14. Tarsals; 15. Metatarsals;
16. Phalanges

Lower Limb
1. Femur; 2. Patella; 3. Tibia; 4. Tarsals;
5. Metatarsals; 6. Phalanges

Articulations
A. 1. Fibrous joint; 2. Cartilaginous joint; 3. Synovial
joint
B. 1. Articular cartilage; 2. Joint cavity; 3. Joint
capsule; 4. Synovial membrane; 5. Bursa
C. 1. Bursa; 2. Articular cartilage; 3. Joint cavity;
4. Joint capsule; 5. Synovial membrane

Types of Synovial Joints
1. Plane joint; 2. Saddle joint; 3. Hinge joint;
4. Pivot joint; 5. Ball-and-socket joint; 6. Ellipsoid
joint

Types of Movement
1. Circumduction; 2. Abduction; 3. Adduction;
4. Flexion; 5. Flexion; 6. Extension; 7. Extension;
8. Flexion; 9. Flexion; 10. Pronation

Joint Disorders
1. Rheumatoid arthritis; 2. Degenerative Joint
Disease (osteoarthritis); 3. Gout; 4. Bursitis;
5. Bunion; 6. Sprain; 7. Separation; 8. Dislocation

1. Osteoblasts, osteocytes, and osteoclasts
2. Compact and cancellous bone
3. Cervical: 7; thoracic: 12; lumbar: 5; sacrum: 1; coccyx: 1
4. True ribs: 7; false ribs: 5; floating ribs: 2
5. Upper limb: carpals 8, metacarpals 5, phalanges 14; lower limb: tarsals 7, metatarsals 5, phalanges 14
6. Fibrous, cartilaginous, and synovial
7. Plane, or gliding joints: articular processes between vertebrae; saddle joints: joint at base of thumb; hinge joints: elbow and knee; pivot joints: between atlas and axis; ball-and-socket joints: shoulder and hip; ellipsoid or condyloid: between occipital condyles and atlas
8. Flexion, extension, abduction, adduction, pronation, supination, rotation, and circumduction

MASTERY LEARNING ACTIVITY

1. E. Bone performs all of the functions listed.

2. E. All of the statements concerning bone matrix are correct.

3. C. The shaft of a long bone is the diaphysis. It is covered by periosteum. Fontanels are the portions of membrane around the brain that are not ossified (converted to bone) at birth. The ends of long bones are the epiphyses; they are covered by articular cartilage.

4. D. Periosteum covers the surface of mature bones, regardless of how they are formed. Articular cartilage covers the ends of long bones. A callus is a fibrous network with cartilage that holds bone fragments together after the bone has been broken.

5. D. Canaliculi ("small canals") are tiny canals that allow cell processes from osteocytes to connect with each other through the lamellae. Canaliculi provide a pathway for nutrients to diffuse to, and for waste products to diffuse away from the osteocytes. Periosteum covers the surface of bones. Haversian canals contain blood vessels, and lacunae are spaces in which the osteocytes are located.

6. C. Areas of membrane around the brain that have not ossified (changed to bone) at birth are called fontanels. Fontanels are found in skull bones, but not in long bones, which are formed from cartilage.

7. B. Osteoclasts break down bone, and osteoblasts build bone. The interaction between these two types of cells is responsible for bone growth and remodeling.

8. E. Following a fracture, bleeding from blood vessels in the bone produces a blood clot. Cells from surrounding tissues invade the clot, and form a fibrous network containing cartilage. This region of repair is called a callus. Then osteoblasts invade the callus and begin producing bone, which replaces the callus.

9. B. Osteomyelitis is inflammation of the bone, often caused by bacterial infection. Osteogenesis imperfecta is a group of genetic disorders in which a lack of collagen produces brittle, easily fractured bones. Rickets is most often a disease of children caused by nutritional deficiencies of vitamin D or calcium. Osteomalacia is softening of the bones resulting from calcium depletion, and osteoporosis, or porous bone, results from reduction in the overall quantity of bone tissue.

10. C. The appendicular skeleton consists of the pectoral (clavicle, scapula) and pelvic (coxa, sacrum) girdles plus the upper and lower limbs. The axial skeleton consists of the skull, vertebrae, and rib cage.

11. D. The vomer and the perpendicular plate of the ethmoid form the bony part of the nasal septum. The nasal bones form the bridge of the nose, and the nasal conchae are three bony shelves in the lateral walls of the nasal cavity. The zygomatic arch is formed from the zygomatic and temporal bones, and forms a bridge across the cheek.

12. C. The temporal bone does not have a paranasal sinus (cavity connected to the nasal cavity).

13. A. The nasolacrimal canal passes from the orbit into the nasal cavity. The optic foramen is the passageway for the optic nerve, the paranasal sinuses are cavities in several of the bones associated with the nasal caviity, and the foramen magnum is the opening through which the spinal cord communicates with the brain.

14. C. There are eight carpal bones in the wrist.

15. E. The joint capsule surrounds and supports the joint. The articular cartilage covers the ends of bones within the joint. The synovial membrane secretes synovial fluid, which lubricates the joint.

16. D. Supination of the forearm turns the knob clockwise, unlatching the door.

17. A. Rheumatoid arthritis is an autoimmune disease that destroys connective tissue, especially the articular cartilage in joints. Degenerative Joint Disease (DJD) is the same thing as osteoarthritis; this condition results from "wear and tear" destruction of articular cartilage in joints. Gout is an inflammatory disease, particularly of the leg and foot, that results from an increase in uric acid in the body. Bursitis is an inflammation of the bursae around joints.

 ☆ **FINAL CHALLENGES** ☆

1. A ruptured disk is most likely to develop in the lumbar region because it supports more weight that the cervical or thoracic regions. The sacral vertebrae are fused.

2. Scoliosis would be expected, because with one lower limb shorter than the other limb, the hips would be abnormally tilted sideways. The vertebral column would curve laterally in an attempt to compensate.

3. The coccyx can bend out of the way (posteriorly) to form a larger opening for the passage of the baby during childbirth.

4. Premature fusion of the cranial bones impairs skull growth and brain growth. The skull may be severely deformed because of the pressure exerted by the growing brain. Surgical procedures to separate the skull bones and allow normal development are usually successful.

5. The statement is true because the bones of the pelvic girdle are strong. If the seat belt is fastened around the abdomen, soft tissues and internal organs are more likely to be damaged.

The Muscular System

FOCUS: Muscle tissue is specialized to contract with a force; it is responsible for body movements. According to the sliding filament mechanism, the movement of actin myofilaments past myosin myofilaments results in the shortening of muscle fibers (cells) and therefore muscles. The three types of muscle tissue are skeletal, smooth, and cardiac muscle.

A muscle that causes a particular movement such as flexing the forearm is called a prime mover. The opposite movement (extending the forearm) is achieved by a different muscle, the antagonist. The study of muscle actions can be approached by examining groups of muscles first and then individual muscles within a group.

CONTENT LEARNING ACTIVITY

Functions of Skeletal Muscle

"Skeletal muscle comprises approximately 40% of the body's weight."

Match these terms with the correct statement or definition:

Heat
Movement
Posture

_____ 1. Result of the contractions of skeletal muscle attached to bones or skin.

_____ 2. Maintenance of body position as a result of continual contraction of skeletal muscle.

_____ 3. Byproduct of the chemical reactions that produce the energy required for muscle contractions.

 Shivering is rapid muscle contractions that produce shaking and result in heat production.

Muscle Structure

Skeletal muscles are composed of skeletal muscle fibers and associated connective tissue.

A. Match these terms with the correct statement or definition:

Actin myofilament
Fascia
Muscle bundle
Muscle fiber

Myofibril
Myosin myofilament
Sarcomere
Striations

_____ 1. Connective tissue sheath that surrounds and separates skeletal muscle.

_____ 2. Group of muscle fibers surrounded by connective tissue.

_____ 3. Single muscle cell.

_____ 4. Thread-like structure that extends from one end of a muscle fiber to the other.

_____ 5. Two proteins that make up a myofibril.

_____

_____ 6. Part of a myofibril extending from one Z line to the next Z line.

_____ 7. Myofilament that attaches to a Z line.

_____ 8. Banding pattern that results from the organization of the actin and myosin myofilaments.

B. **M**atch these terms with
the correct parts labeled
in Figure 7-1:

Fascia
Muscle bundles
Muscle fibers
Myofibrils
Myofilaments
Sarcomere

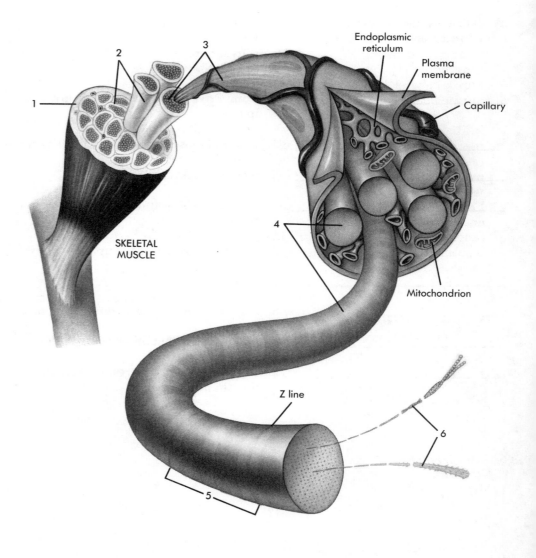

1. _____   3. _____   5. _____

2. _____   4. _____   6. _____

Muscle Contraction

"*Muscle contraction is the forceful shortening of a muscle.***"**

A. Using the terms provided, complete the following statements:

Actin myofilaments   Myofibrils
Cross bridge   Myosin myofilaments
Muscle bundles   Z lines
Muscle fibers

A small connection from the myosin myofilament to the actin myofilament is called a _(1)_. Movement of the cross bridges cause _(2)_ to move past _(3)_. This is called the sliding filament mechanism. It causes the _(4)_ of sarcomeres to move closer together, resulting in shortening of the sarcomeres. Consequently, _(5)_, which are composed of sarcomeres joined end-to-end, also shorten. The shortening of the myofibrils causes _(6)_ to shorten, which in turn causes _(7)_ to shorten. The whole muscle shortens as a result of the shortening of the muscle bundles.

1. _____
2. _____
3. _____
4. _____
5. _____
6. _____
7. _____

☞ Relaxation occurs when cross bridges release actin myofilaments which return to their original precontraction position as another contracting muscle or gravity pulls on the relaxed muscle.

B. Match these terms with the correct parts labeled in Figure 7-2:

Actin myofilament
Cross bridge
Myosin myofilament
Sacromere
Z line

1. _____   3. _____   5. _____

2. _____   4. _____

C. Match these terms with the correct statement or definition:

Isometric
Isotonic
Muscle tone
Recruitment

_____ 1. An increase in the number of muscle fibers contracting.

_____ 2. Contraction in which the amount of tension is constant, but the length of the muscle changes.

_____ 3. Contraction in which the length of muscle does not change, but the amount of tension increases.

_____ 4. Responsible for movements of the fingers to make a fist.

_____ 5. Clenching the fist harder and harder is an example.

_____ 6. Constant tension produced for long periods of time; responsible for maintaining posture.

D. Match these terms with the correct statement or definition:

Aerobic respiration
Anaerobic respiration
ATP
Oxygen debt
Muscle fatigue
Psychological fatigue

_____ 1. Molecule used to provide energy for muscle contraction.

_____ 2. Type of respiration that requires oxygen.

_____ 3. Type of respiration that produces lactic acid.

_____ 4. Type of respiration that produces the most ATP for each glucose molecule used.

_____ 5. Type of respiration that uses fatty acids to generate ATP.

_____ 6. Type of respiration used during short periods of intense exercise.

_____ 7. Amount of oxygen needed to convert lactic acid to glucose.

_____ 8. Results when ATP is used during muscle contraction faster than it can be produced in muscle cells, and lactic acid builds up faster than it can be removed.

_____ 9. The reason most people stop exercising.

Understanding Muscle Fibers

66 *Muscles fibers are classified as either fast-twitch or slow-twitch muscle fibers.* **99**

Match these terms with the
correct statement or definition:

Atrophy   Fast-twitch muscle fibers
Hypertrophy   Slow-twitch muscle fibers

_____ 1. Most resistant to fatigue.

_____ 2. Has a richer blood supply and contains myoglobin, which
temporarily stores oxygen.

_____ 3. White meat of a chicken's breast is an example.

_____ 4. Increase in the size of a muscle resulting from an increase in the
number of myofibrils within muscle fibers.

_____ 5. Intense exercise resulting in anaerobic respiration has the greatest
effect on this type of muscle fiber.

General Principles of Muscle Anatomy

66 *Muscle contraction causes body movements by pulling one bone toward another across a movable joint.* **99**

Match these terms with the
correct statement or definition:

Antagonist   Origin
Insertion   Synergists
Prime mover   Tendon

_____ 1. Attaches a muscle to a bone.

_____ 2. End of the muscle attached to the bone undergoing the greatest
movement.

_____ 3. Muscles that work together to accomplish a movement.

_____ 4. Muscle that plays the major role in accomplishing a particular
movement.

_____ 5. Muscle working in opposition to another muscle.

Muscles of the Head and Neck

"*Head muscles are responsible for facial expression and mastication. Neck muscles move the head.*"

A. Match these terms with the correct statement or definition:

Buccinator   Occipitofrontalis
Corrugator supercilii   Orbicularis oculi
Depressor anguli oris   Orbicularis oris
Levator labii superioris   Zygomaticus

_____ 1. Raises the eyebrows.

_____ 2. Closes the eyelids.

_____ 3. Two muscles that pucker the lips

_____

_____ 4. Responsible for smiling.

_____ 5. Responsible for sneering.

_____ 6. Two muscles responsible for frowning.

_____

B. Match these terms with the correct statement or definition:

Masseter   Temporalis
Sternocleidomastoid   Trapezius

_____ 1. Easily seen and felt on the side of the head, these two muscles close the jaw.

_____

_____ 2. Flexes the neck and rotates the head.

_____ 3. Extends the head and neck

Trunk Muscles

"Trunk muscles include those of the vertebral column and those of the abdominal wall."

Match these terms with the correct statement or definition:

Deep back muscles
Erector spinae
External abdominal oblique

Internal abdominal oblique
Rectus abdominis
Transversus abdominis

_____ 1. Group of muscles on each side of the back; primarily responsible for keeping the back straight; also extend, abduct, and rotate the vertebral column.

_____ 2. Muscles located between the spinous and transverse processes of adjacent vertebrae.

_____ 3. Anterior muscle located on each side of the midline; flexes the vertebral column.

_____ 4. Most deep lateral abdominal wall muscle.

☞ The muscle bundles of the abdominal muscles are oriented in different directions, forming a strong abdominal wall that can produce different movements of the vertebral column.

Upper Limb Muscles

"The muscles of the upper limb include those that attach the limb and girdle to the body and those that are in the arm, forearm, and hand."

A. Match these terms with the correct statement or definition:

Deltoid
Latissimus dorsi
Pectoralis major

Serratus anterior
Trapezius

_____ 1. Attaches the scapula to the thorax; forms the upper line from each shoulder to the neck.

_____ 2. Attaches the scapula to the thorax; anterior chest muscle.

_____ 3. Attaches the humerus to the thorax; flexes and adducts the arm.

_____ 4. Attaches the humerus to the thorax; extends and adducts the arm.

_____ 5. Attaches the humerus to the scapula; major abductor of the upper limb.

☞ The muscles that attach the scapula to the thorax hold the scapula in position when the arm muscles contract. They also move the scapula into different positions, thereby increasing the range of movement of the upper limb.

B. Match these terms with the correct statement or definition:

Anterior forearm muscles   Intrinsic hand muscles
Biceps brachii   Posterior forearm muscles
Brachialis   Triceps brachii

_____ 1. Posterior arm muscle; extends the forearm.

_____ 2. Two anterior arm muscles; flex the forearm.

_____

_____ 3. Group of muscles that flex the wrist and fingers.

_____ 4. Group of muscles that extend the wrist and fingers.

_____ 5. Muscles located in the hand; responsible for finger and thumb movements.

Lower Limb Muscles

"_The lower limb muscles include those located in the hip, thigh, leg, and foot._**"**

A. Match these terms with the correct statement or definition:

Anterior thigh muscles   Iliopsoas
Gluteus maximus   Medial thigh muscles
Gluteus medius   Posterior thigh muscles

_____ 1. Anterior hip muscle that flexes the thigh.

_____ 2. Forms most of the mass of the buttocks; extends and abducts the thigh.

_____ 3. Common site for injections; abducts the thigh.

_____ 4. Group of thigh muscles that flex the thigh.

_____ 5. Group of thigh muscles that extend the thigh.

_____ 6. Group of thigh muscles that adduct the thigh.

B. **M**atch these terms with the
correct statement or definition:

Anterior leg muscles	Peroneus
Gastrocnemius	Quadriceps femoris
Hamstring muscles	Sartorius
Intrinsic foot muscles	

_____ 1. Anterior thigh muscle; extends the leg; sometimes used as an injection site.

_____ 2. Anterior thigh muscle; flexes the leg.

_____ 3. Posterior thigh muscle; flexes the leg.

_____ 4. Leg muscle that joins the calcaneal (Achilles) tendon; plantar flexes the foot.

_____ 5. Leg muscles that dorsiflex the foot and extend the toes.

_____ 6. Lateral leg muscles that evert the foot.

_____ 7. Muscles located in the foot; responsible for toe movements.

Disorders of Muscle Tissue

66_Understanding muscle disorders is clinically important._**99**

Match these terms with the
correct statement or definition:

Cramps	Muscular dystrophy
Denervation	Myasthenia gravis
Fibromyalgia	Polio
Multiple sclerosis	Tendonitis

_____ 1. Interruption of the nerve supply to a muscle; results in flaccid paralysis and atrophy.

_____ 2. Muscular weakness and fatigue that occurs when the immune system attacks the connection between muscle fibers and nerve cells.

_____ 3. Viral disease that destroys the nerve cells supplying skeletal muscles, resulting in paralysis.

_____ 4. Usually inherited disease that results in degeneration of muscle tissue and replacement of the muscle tissue with connective tissue.

_____ 5. Painful, spasmodic contractions of muscles that are usually the result of an irritation within a muscle.

_____ 6. Inflammation of a tendon and/or its attachment site.

_____ 7. Widespread, consistent aches within muscles or where the muscles join tendons; cause unknown.

Location of Superficial Muscles

A. Match these terms with the correct parts labeled in Figure 7-3:

Adductor muscles
Anterior forearm muscles
Biceps brachii
Deltoid
External abdominal oblique
Gastrocnemius
Pectoralis major
Quadriceps femoris
Rectus abdominis
Sartorius
Serratus anterior
Sternocleidomastoid
Tibialis anterior

1. _____

2. _____

3. _____

4. _____

5. _____

6. _____

7. _____

8. _____

9. _____

10. _____

11. _____

12. _____

13. _____

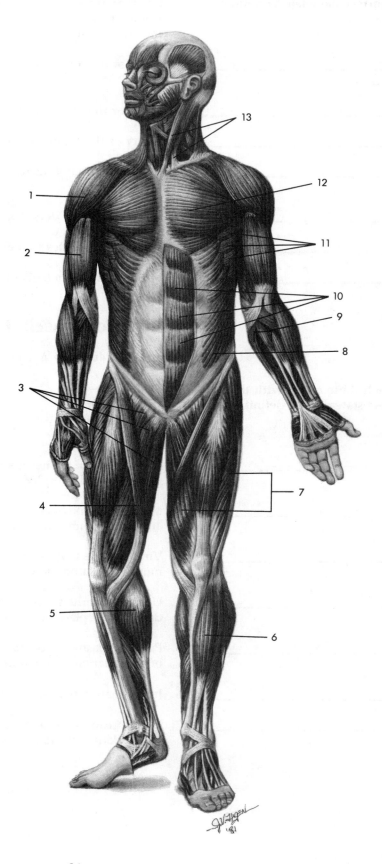

B. Match these terms with
the correct parts labeled
in Figure 7-4:

Calcaneal tendon
Deltoid
Gastrocnemius
Gluteus maximus
Hamstrings
Latissimus dorsi
Posterior forearm muscles
Trapezius
Triceps brachii

1. _____

2. _____

3. _____

4. _____

5. _____

6. _____

7. _____

8. _____

9. _____

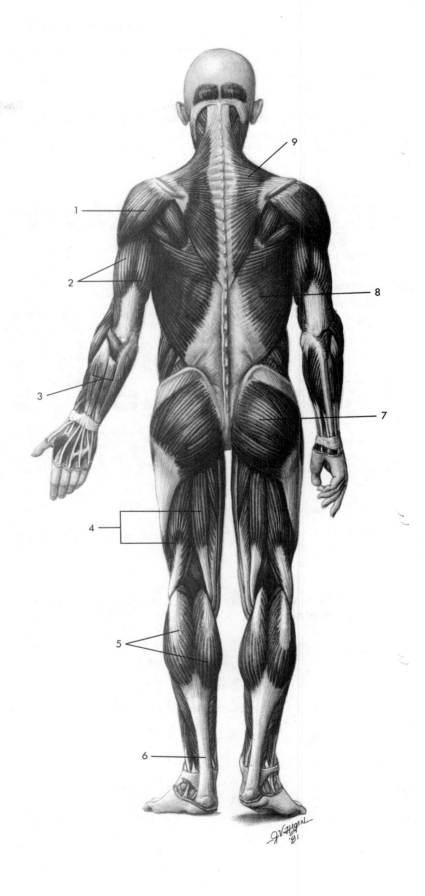

1. List the three major functions of skeletal muscle.

2. Arrange the following structures in order, from the outside to the inside of a skeletal muscle: muscle bundle, muscle fiber, myofibril, myofilament.

3. List the events that result in the shortening of a contracting muscle.

4. Explain how the force of muscle contractions can vary from weak contractions to very strong contractions.

5. Name two types of muscle contraction.

6. State the two way by which skeletal muscles produce the energy (ATP) necessary for muscle contraction.

7. Name two types of skeletal muscle fibers.

MASTERY LEARNING ACTIVITY

Place the letter corresponding to the correct answer in the space provided.

_____ 1. Fascia
a. is the connective tissue sheath surrounding and separating muscles.
b. are skeletal muscles responsible for facial expression.
c. attaches skeletal muscles to the skin.
d. is primarily responsible for the production of the heat necessary to maintain body temperature.

_____ 2. Given the following structures:
1. muscle bundle
2. muscle fiber
3. myofibril
4. myofilament

Choose the arrangement that lists the structures in the correct order from the inside to the outside of a skeletal muscle.
a. 3, 4, 1, 2
b. 3, 4, 2, 1
c. 4, 2, 3, 1
d. 4, 3, 2, 1

_____ 3. Sarcomeres
a. contain myofilaments.
b. extend from one Z line to the next Z line.
c. joined end-to end form myofibrils.
d. all of the above

_____ 4. Myosin myofilaments
a. are attached to the Z line.
b. form cross bridges with actin myofilaments.
c. are thinner than actin myofilaments.
d. all of the above

_____ 5. Given the following events:
1. myosin combines with actin
2. repeated cross bridge movements
3. actin slides past myosin
4. another contracting muscle or gravity pulls on the relaxed muscle

Following a contraction a muscle is shortened, and as it relaxes muscle length increases. Which of the events listed result in increased length of a relaxed muscle?
a. 4
b. 1, 2
c. 3, 4
d. 1, 2, 3

_____ 6. Given the following events:
1. involuntary reflex contractions
2. muscle tone
3. recruitment
4. voluntary control of contractions

Which of the events result from the excitable nature of skeletal muscles?
a. 1, 4
b. 2, 3
c. 1, 2, 3
d. 2, 3, 4
e. 1, 2, 3, 4

_____ 7. A weight-lifter attempts to lift a weight from the floor, but the weight is so heavy he is unable to move it. The contractions he used was mostly
a. isometric.
b. isotonic.
c. notsometric.
d. notsotonic.
e. notsoeasy.

_____ 8. Contrasting aerobic and anaerobic respiration,
a. anaerobic respiration produces more ATP per glucose molecule than does aerobic respiration.
b. anaerobic respiration requires oxygen, whereas aerobic respiration does not requires oxygen.
c. anaerobic respiration produces lactic acid, and aerobic respiration produces carbon dioxide and water.
d. anaerobic respiration produces ATPs slowly compared to aerobic respiration.

9. Dudly Smartlips pulled into Kentucky Fried Chicken and placed an order for a McChicken sandwich. But, he wanted low myoglobin meat, hold the mayo. He was served
   a. chicken breasts (white meat)
   b. chicken thighs (dark meat)

10. Fast-twitch fibers
    a. can be changed into slow-twitch fibers with exercise.
    b. are found in higher proportions in arm muscles than in back muscles.
    c. are found in higher proportions in the thigh muscles of marathon runners than in the thigh muscles of sprinters.
    d. have more mitochondria than slow-twitch fibers

11. Muscles that oppose one another are
    a. synergist.
    b. hateful.
    c. prime movers.
    d. antagonists.

12. An aerial circus performer who supports herself only by her teeth while spinning around and around should have strong
    a. temporalis and masseter muscles.
    b. zygomaticus muscles.
    c. trapezius muscles.
    d. orbicularis oris muscles.

13. A man lies flat on his back. While someone holds his feet he does a "sit-up." Which of the following muscles would be involved?
    a. rectus abdominis
    b. iliopsoas
    c. anterior thigh muscles
    d. all of the above

14. Which of the following muscles would one expect to be especially well developed in a boxer whose specialty is a "jab?"
    a. biceps brachii
    b. brachialis
    c. deltoid
    d. triceps brachii

15. Which of the following would be well developed in a football player whose specialty is kicking field goals?
    a. hamstrings
    b. quadriceps femoris
    c. gluteus maximus
    d. gastrocnemius

16. Which of the following muscles would be especially well developed in a ballerina?
    a. pectoralis major
    b. rectus abdominis
    c. gastrocnemius
    d. erector spinae

FINAL CHALLENGES

Use a separate sheet of paper to complete this section.

1. The following experiments were performed in an anatomy and physiology laboratory. The rate and depth of respiration for a resting student was determined. In experiment A the student ran in place for 30 seconds, immediately sat down and relaxed, and respiration rate and depth was again determined. Experiment B was just like experiment A except that the student held her breath while running in place. What differences in respiration would you expect for the two different experiments. Explain the basis for your predictions.

2. Sally Gorgeous, an avid jogger, is running down the beach when she meets Sunny Beachbum, an avid weight lifter. Sunny flirts with Sally, who decides he has more muscles than brains. She runs down the beach, but Sunny runs after her. After about a half mile Sunny tires and gives up. Explain why Sally was able to outrun Sunny (i.e., do more muscular work) despite the fact that she obviously is less muscular.

3. Describe the movement each of these muscles produces: biceps brachii, hamstrings, and pectoralis major. Name the muscles that act as synergists and antagonists for each movement of the muscle.

4. According to an advertisement, a special type of sandal makes the calf look better because the calf muscles are exercised when the toes curl and grip the sandal while walking. Explain why you believe or disbelieve this claim.

5. According to an advertisement, the appearance of the thighs can be improved by exercising the thigh muscles. While in a sitting position, the exercise device is placed between the knees, and the knees are repeatedly squeezed together against the resistance of the exercise device. Explain why you believe or disbelieve this claim.

ANSWERS TO CHAPTER 7

CONTENT LEARNING ACTIVITY

Functions of Skeletal Muscle
1. Movement;. 2. Posture; 3. Heat
Muscle Structure
A. 1. Fascia; 2. Muscle bundle; 3. Muscle fiber; 4. Myofibril; 5. Actin myofilament and myosin myofilament; 6. Sarcomere; 7. Actin myofilament; 8. Striations
B. 1. Fascia; 2. Muscle bundles; 3. Muscle fibers; 4. Myofibrils; 5. Sarcomere; 6. Myofilaments
Muscle Contraction
A. 1. Cross bridge; 2. Actin myofilaments; 3. Myosin myofilaments; 4. Z lines; 5. Myofibrils; 6. Muscle fibers; 7. Muscle bundles
B. 1. Z line; 2. Actin myofilament; 3. Myosin myofilament; 4. Cross bridge; 5. Sarcomere
C. 1. Recruitment; 2. Isotonic; 3. Isometric; 4. Isotonic; 5. Isometric; 6. Muscle tone
D. 1. ATP; 2. Aerobic respiration; 3. Anaerobic respiration; 4. Aerobic respiration; 5. Aerobic respiration; 6. Anaerobic respiration; 7. Oxygen debt; 8. Muscle fatigue; 9. Psychological fatigue
Understanding Muscle Fibers
1. Slow-twitch muscle fibers; 2. Slow-twitch muscle fibers; 3; Fast-twitch muscle fibers; 4. Hypertrophy; 5. Fast-twitch muscle fibers
General Principles of Muscle Anatomy
1. Tendon; 2. Insertion; 3. Synergist; 4. Prime mover; 5. Antagonist
Muscles of the Head and Neck
A. 1. Occipitofrontalis; 2. Orbicularis oculi; 3. Orbicularis oris and buccinator; 4. Zygomaticus; 5. Levator labii superioris; 6. Depressor anguli oris and corrugator supercilii
B. 1. Temporalis and masseter; 2. Sternocleidomastoid; 3. Trapezius

Trunk Muscles
1. Erector spinae; 2. Deep back muscles; 3. Rectus abdominis; 4. Transversus abdominis
Upper Limb Muscles
A. 1. Trapezius; 2. Serratus anterior; 3. Pectoralis major; 4. Latissimus dorsi; 5. Deltoid
B. 1. Triceps brachii; 2. Biceps brachii and brachialis; 3. Anterior forearm muscles; 4. Posterior forearm muscles; 5. Intrinsic hand muscles
Lower Limb Muscles
A. 1. Iliopsoas; 2. Gluteus maximus; 3. Gluteus medius; 4. Anterior thigh muscles; 5. Posterior thigh muscles; 6. Medial thigh muscles
B. 1. Quadriceps femoris; 2. Sartorius; 3. Hamstring muscles; 4. Gastrocnemius; 5. Anterior leg muscles; 6. Peroneus; 7. Intrinsic foot muscles
Disorders of Muscle Tissue
1. Denervation; 2. Myasthenia gravis; 3. Polio; 4. Muscular dystrophy; 5. Cramps; 6. Tendonitis; 7. Fibromyalgia
Location of Superficial Muscles
A. 1. Deltoid; 2; Biceps brachii; 3. Adductor muscles; 4. Sartorius; 5. Gastrocnemius; 6. Tibialis anterior; 7. Quadriceps femoris; 8. External abdominal oblique; 9. Anterior forearm muscles; 10. Rectus abdominis; 11. Serratus anterior; 12. Pectoralis major; 13. Sternocleidomastoid
B. 1. Deltoid; 2. Triceps brachii; 3. Posterior forearm muscles; 4. Hamstrings; 5. Gastrocnemius; 6. Calcaneal tendon; 7. Gluteus maximus; 8. Latissimus dorsi; 9. Trapezius

1. Movement, posture, and heat production
2. Muscle bundle, muscle fiber, myofibril, myofilament
3. Myosin myofilaments bind to actin myofilaments forming cross bridges. The cross bridges move, causing actin myofilaments to slide past myosin myofilaments. Consequently, Z lines move closer together and sarcomeres shorten. Shortening of sarcomeres results in shortening of myofibrils, then muscle fibers, then muscle bundles, then the whole muscle

4. Weak force of contraction results from few muscle fibers contracting, whereas a stronger force of contraction results from a greater number of muscle fibers contracting. The increase in the number of muscle fibers contracting is called recruitment
5. Isotonic and isometric contractions
6. Anaerobic and aerobic respiration
7. Slow-twitch and fast-twitch muscle fibers

MASTERY LEARNING ACTIVITY

1. A. Fascia is a connective tissue sheath that surrounds and separates muscles. The hypodermis attaches skeletal muscle to skin. Muscle is primarily responsible for heat production.

2. D. Actin and myosin myofilaments makeup myofibrils within muscle fibers. Muscle fibers makeup muscle bundles.

3. D. Sarcomeres extend from one Z line to the next Z line. Actin and myosin myofilaments are in-between each Z line. Sarcomeres joined end-to-end form myofibrils, and muscle bundles makeup muscles.

4. B. Myosin myofilaments have extensions that attach to actin myofilaments to form cross bridges. Actin myofilaments are attached to the Z line, and myosin myofilaments are thicker than actin myofilaments.

5. C. Relaxation occurs when myosin myofilaments stop forming cross bridges with actin myofilaments. The muscle lengths as actin myofilaments slide past myosin myofilaments because another contracting muscle or gravity pulls on the relaxed muscle.

6. E. Muscles are excitable, which means they respond to stimuli by contracting. Normally the skeletal muscles are stimulated by the nervous system. All the events listed result from nervous system stimulation.

7. A. A contraction in which the length of the muscle does not change is isometric. Because the weight did not move, the weight-lifter's limbs did not change position, and therefore neither did the length of his muscles.

8. C. Anaerobic respiration produces lactic acid, and aerobic respiration produces carbon dioxide and water. Anaerobic respiration does not require oxygen, produces fewer ATP per glucose molecule than aerobic respiration, but produces them faster.

9. A. Low myoglobin content would be found in muscle that was predominantly composed of fast-twitch fibers. Such muscles can contract rapidly, but fatigue quickly. Chickens can fly only short distances. On the other hand, dark meat with high myoglobin content is typical of muscle predominantly composed of slow-twitch fibers. Such muscles are fatigue resistant; an example is chicken thighs.

10. B. Fast-twitch fibers are more abundant in the arms, which are capable of quick movements, than the back, which are fatigue resistant and maintain posture. A marathon runner would have a higher proportion of slow-twitch fibers than a sprinter. Exercise does not cause one type of fiber to switch to another type. However, exercise can make fast-twitch fibers more fatigue resistant.

11. D. Two or more muscles that oppose one another are antagonists. Muscles that assist each other are synergists. A prime mover is the muscle mostly responsible for a movement.

12. A. The temporalis and masseter muscles close the jaw

13. D. The iliopsoas is responsible for most of the flexion that occurs between the trunk and thigh. The anterior thigh muscles also contribute. The rectus abdominis causes flexion of the vertebral column.

14. D. Extension of the forearm during a punch is critical to a boxer; the triceps brachii performs that function.

15. B. Contraction of the quadriceps femoris can cause the leg to be forcefully extended, as in kicking a football.

16. C. Because a ballerina often dances on her toes, plantar flexion of the foot is necessary. The gastrocnemius muscles perform that function.

☆ FINAL CHALLENGES ☆

1. In experiment A the student used anaerobic respiration as she started to run in place, but aerobic respiration also increased to meet most of her energy needs. When she stopped running, respiration rate would be increased over resting levels because of repayment of the oxygen debt as a result of the anaerobic respiration. In experiment B almost all of her energy would come from anaerobic respiration because she is holding her breath while running in place. Consequently, she would have a much larger oxygen debt. One would predict that following running in place in experiment B her respiration rate would be greater than in experiment A, or that her respiration rate would be elevated for a longer period of time than in experiment A, or both.

2. Sally's aerobic exercise program of jogging has developed her slow-twitch muscle fibers and increased the fatigue resistance of her fast-twitch muscle fibers. Sunny's weight-lifting program consists of intense, but short, periods of exercise. This increases strength by increasing the size of his muscle fibers. Because it relies on anaerobic respiration, however, it does not develop aerobic, fatigue-resistance abilities. Sunny needs to do some aerobic exercises, which are included in most modern day weight-lifting programs.

3. The biceps brachii flexes the forearm. Its synergist is the brachialis (also flexes the forearm) and its antagonist is the triceps brachii (extends the forearm).

   The hamstrings extend the thigh and flex the leg. Its synergists are the gluteus maximus (extend the thigh) and sartorius (flex the leg); its antagonists are the iliopsoas (flex the thigh) and quadriceps femoris (extend the leg).

   The pectoralis major adducts and flexes the arm. Its synergists are the latissimus dorsi (adducts the arm) and deltoid (flexes the arm); its antagonists are the deltoid (abducts the arm) and latissimus dorsi (extends the arm).

4. Disbelieve this advertisement. The deep posterior leg muscles flex the toes and are covered by the gastrocnemius. Even if the sandals cause some enlargement of the deep posterior leg muscles, it would not be noticeable. To improve the appearance of the calf, the gastrocnemius should be exercised. This muscle attaches to the calcaneus and causes plantar flexion of the foot. It does not cause the toes to be flexed, so flexing the toes would have no effect on the gastrocnemius.

5. The exercise device develops the medial thigh muscles, which cause adduction of the thigh. Because the device does not exercise the anterior or posterior thigh muscles, used alone it would not provide a way to develop the entire thigh.

101

The Nervous System

FOCUS: The nervous system can be divided into the central nervous system (brain and spinal cord) and the peripheral nervous system (receptors, nerves, and ganglia). Cells of the nervous system are neurons, which conduct action potentials, and neuroglia, which provide support and nourishment. Synapses are connections between neurons. The major regions of the brain are the brainstem, diencephalon, cerebrum and cerebellum. The spinal cord connects peripheral nerves to the brain. The peripheral nervous system consists of sensory and motor neurons. Motor neurons can be subdivided into the somatic and autonomic systems. The autonomic system can be divided into sympathetic and parasympathetic divisions.

CONTENT LEARNING ACTIVITY

Divisions of the Nervous System

66*The nervous system can be divided into the central and peripheral nervous systems.*99

Match these terms with the correct statement or definition:

Autonomic nervous system   Peripheral nervous system
Central nervous system   Sensory division
Motor division   Somatic nervous system

_____ 1. Consists of the brain and spinal cord.

_____ 2. Consists of receptors, nerves, and ganglia located outside the CNS.

_____ 3. PNS division that transmits action potentials from sensory organs to the CNS.

_____ 4. Subdivision of the motor division that transmits action potentials from the CNS to skeletal muscle only.

_____ 5. Subdivision of the motor division that transmits action potentials to smooth muscle, cardiac muscle, or glands.

Cells of the Nervous System

"_Cells of the nervous system are neurons and neuroglia._**"**

A. Match these terms with the correct statement or definition:

Axons	Dendrites
Cell body	Nerve fibers

_____ 1. Location of the nucleus and most protein synthesis.

_____ 2. Usually carry electrical signals to the cell body.

_____ 3. Usually carry electrical signals away from the cell body.

_____ 4. General name for dendrites and axons.

B. Match these terms with the correct statement or definition:

Association neurons	**Sensory neurons**
Motor neurons	

_____ 1. Conduct action potentials to the CNS.

_____ 2. Conduct action potentials away from the CNS.

_____ 3. Located primarily within the CNS.

C. Using the terms provided, complete the following statements:

Blood-brain barrier	Increase
Decrease	Myelin sheaths
Glioma	Nodes of Ranvier

1. _____

2. _____

3. _____

4. _____

5. _____

The most common brain tumor, called a (1) , is actually formed from neuroglia cells, not from neurons. Neuroglia cells form the (2) around axons. The myelin sheaths insulate the axons and (3) the speed at which action potentials travel along an axon. In-between the neuroglia cells are small bare areas of the axon called (4) . Some neuroglia cells surround capillaries and are responsible for the (5) , which protects the brain and spinal cord from harmful substances.

D. Match these terms with the correct parts labeled in Figure 8-1:

Axon
Cell body
Dendrite
Neuroglia cell
Node of Ranvier

1. _____

2. _____

3. _____

4. _____

5. _____

Nucleus

Nissl bodies

Organization of Nervous Tissue

" *Bundles of axons form nerve tracts or nerves, whereas cell bodies and their dendrites* **"**
form cortex, nuclei, or ganglia.

Match these terms with the correct statement or definition:

Cortex
Ganglion
Gray matter
Nerve tracts

Nerves
Nuclei
White matter

_____ 1. Color produced by groups of nerve cell bodies and their dendrites.

_____ 2. Gray matter on the surface of the brain.

_____ 3. Clusters of gray matter located deeper within the brain.

_____ 4. Cluster of nerve cell bodies in the PNS.

_____ 5. Color produced by bundles of axons with their myelin sheaths.

_____ 6. Conduction pathways composed of white matter in the CNS.

_____ 7. In the PNS, bundles of nerve fibers and their connective tissue sheaths.

Action Potentials

" *Action potentials move along the membrane of a nerve fiber.* **"**

Using the terms provided, complete the following statements:

Action potential
Myelinated
Negatively charged
Nerve impulse

Node of Ranvier
Positively charged
Resting membrane potential
Unmyelinated

The outside of an unstimulated, resting cell membrane is
 (1) compared with the inside of the cell membrane. This
condition is called the (2) . A brief reversal of these charges in
response to a stimulus such as chemicals, temperature, pressure,
or light , is called a(n) (3) . The movement of the action
potential along the membrane of a nerve fiber is called
a(n) (4) . Conduction of an action potential is more rapid
in (5) fibers than (6) fibers. In myelinated fibers, action
potentials are only produced at the uninsulated   (7) , so action
potentials jump from one node of Ranvier to the next.

1. _____

2. _____

3. _____

4. _____

5. _____

6. _____

7. _____

Understanding Action Potentials

" *Awareness of stimuli begins with the production of action potentials.* **"**

Match these terms with the
correct statement or definition:

Action potential
All-or-none
Depolarization
K^+ ions
Na^+ ions

Repolarization
Sodium-potassium exchange
pump
Threshold

_____ 1. These ions are present in higher concentration inside the cell than outside the cell.

_____ 2. These ions diffuse into the cell when the cell is stimulated.

_____ 3. Process during which the inside of the cell membrane becomes positive compared to the outside.

_____ 4. Process during which the cell membrane returns to its resting membrane potential.

_____ 5. Depolarization and repolarization together.

_____ 6. Continuously transports Na^+ ions out of the cell and K^+ ions into the cell.

_____ 7. A stimulus produces an action potential if the stimulus is above a certain level; if the stimulus is below that level, no action potential occurs.

_____ 8. The stimulus level required to produce an action potential.

The Synapse

"A synapse is a junction where action potentials can be transferred from one cell to another cell."

Match these terms with the correct statement or definition:

Acetylcholine and norepinephrine   Neurotransmitters
Acetylcholinesterase   Synaptic cleft

_____ 1. Space separating the end of the axon of the neuron from the muscle fiber.

_____ 2. General term for chemicals found in small vesicles in the axon ending; these are released when an action potential reaches the axon ending.

_____ 3. Two common neurotransmitters.

_____ 4. An enzyme that breaks down acetylcholine.

Central Nervous System

"The central nervous system consists of the brain and spinal cord."

Match these terms with the correct statement or definition:

Brain
Spinal cord

_____ 1. Part of the central nervous system housed within the braincase.

_____ 2. Part of the central nervous system surrounded by the vertebral column.

_____ 3. Part of the central nervous system containing the brainstem, diencephalon, cerebrum and cerebellum.

Brainstem

"The brainstem connects the spinal cord to the remainder of the brain and contains many ascending and descending nerve tracts."

Match these terms with the correct statement or definition:

Medulla oblongata   Pons
Midbrain   Reticular activating system

_____ 1. Most inferior part of the brainstem; regulates heart rate, breathing, swallowing, coughing and sneezing.

_____ 2. Part of the brainstem superior to the medulla oblongata; relays information between the cerebrum and cerebellum.

_____ 3. Smallest and most superior region of the brainstem; contains four mounds of tissue involved with hearing and vision.

_____ 4. Nuclei scattered throughout the brainstem that play a role in arousing and maintaining consciousness.

Diencephalon

❝*The diencephalon is the part of the brain between the brainstem and cerebrum.***❞**

A. Match these terms with the correct statement or definition:

Hypothalamus
Thalamus

_____ 1. Largest part of the diencephalon; processes most sensory input from the brainstem.

_____ 2. Contains several small nuclei important in maintaining homeostasis; plays a central role in control of temperature, hunger and thirst; involved in emotional responses to odors.

_____ 3. Connected to the pituitary gland and regulates the release of hormones from the pituitary gland.

B. Match these terms with the correct parts labeled in Figure 8-2:

Hypothalamus
Pituitary gland
Thalamus

1. _____

2. _____

3. _____

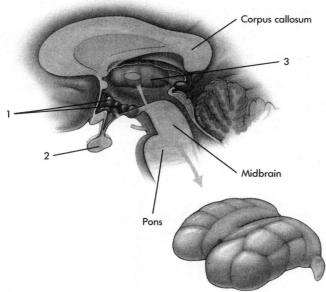

Corpus callosum

3

1

2

Midbrain

Pons

Cerebrum

❝*The cerebrum is the largest portion of the brain.***❞**

A. Match these terms with the correct statement or definition:

Gyri
Sulci

_____ 1. Raised folds on the surface of each cerebral hemisphere.

_____ 2. Grooves found between raised folds on the surface of the cerebrum.

B. Match these lobes of the cerebrum with the correct primary function:

Frontal lobe   Parietal lobe
Occipital lobe   Temporal lobe

_____ 1. Voluntary motor function, motivation, aggression, and mood.

_____ 2. Reception and evaluation of most sensory information, such as pain, temperature, pressure (touch), and taste.

_____ 3. Reception and integration of visual input.

_____ 4. Evaluation of auditory and olfactory input; also memory, abstract thought and judgment.

Cerebral Cortex

“*The gray matter on the outer surface of the cerebrum is the cortex.***”**

Match these terms with the correct statement or definition:

Aphasia   Motor area
Broca's area   Prefrontal area
General sensory area

_____ 1. Specific region of the parietal lobe where sensory input such as pain, temperature and pressure occurs.

_____ 2. Area of the frontal lobe that initiates action potentials for voluntary control of skeletal muscles.

_____ 3. Part of the motor area; responsible for producing the muscle movements necessary for speech.

_____ 4. Absent or defective speech.

_____ 5. Motivation and foresight to plan and initiate movements occur in this area of the anterior portion of the frontal lobe.

☞ Electrodes placed on a person's scalp and attached to a recording device can record the brain's electrical activity, producing an electroencephalogram (EEG). The EEG displays wavelike patterns (brain waves).

Memory

“*Memory is the mental ability to recall information.***”**

Match these terms with the correct statement or definition:

Long-term memory   Short-term memory
Memory engrams

_____ 1. Sensory information that is held for a few seconds to a few minutes.

_____ 2. Memory that may become permanent.

_____ 3. A series of neurons involved in long-term retention of information.

108

Right and Left Cerebral Hemispheres

"_Some functions are not shared equally between the two hemispheres._**"**

Match the hemisphere with the correct function:

Corpus callosum
Left cerebral hemisphere

Right cerebral hemisphere

_____ 1. Controls muscular function in and receives sensory input from the left half of the body.

_____ 2. Largest connection between the two cerebral hemispheres.

_____ 3. Thought to be the analytic hemisphere; emphasizes such skills as mathematics and speech.

Basal Ganglia and Cerebellum

"_The basal ganglia and cerebellum are important in controlling body movements._**"**

Match these terms with the correct statement or definition:

Basal ganglia
Cerebellum

_____ 1. Group of nuclei in the cerebrum that decrease muscle tone and inhibit muscular activity.

_____ 2. Functions as a comparator; involved in balance, maintenance of muscle tone, and fine motor movement.

_____ 3. Capable of "learning" motor skills.

Spinal Cord

"_The spinal cord extends from the foramen magnum to the second lumbar vertebra._**"**

A. Match these terms with the correct statement or definition:

Dorsal root
Dorsal root ganglia
Gray matter

Spinal nerve
Ventral root
White matter

_____ 1. Central portion of the spinal cord; consists of nerve cell bodies and dendrites that synapse with fibers from spinal nerves or the brain.

_____ 2. Peripheral spinal cord area for ascending or descending nerve tracts formed by association neurons.

_____ 3. Sensory nerve fibers where they enter the spinal cord.

_____ 4. Structures containing the cell bodies of sensory neurons.

_____ 5. Formed by dorsal and ventral roots; contains both sensory and motor nerve fibers.

_____ 6. Nerve fibers of motor neurons where they leave the spinal cord.

B. Match these terms with the correct parts labeled in Figure 8-3:

Dorsal root   Ventral root
Dorsal root ganglion   White matter
Gray matter

1. __ _____
2. _____
3. _____
4. _____
5. _____

☞ For most ascending and descending nerve tracts, the nerve fibers cross from one side of the body to the other side. Because of this cross over, one side of the brain receives input from and controls the opposite side of the body.

Reflexes

"A reflex is a response to a stimulus that does not involve conscious thought."

A. Using the terms provided, complete the following statements:

Association neurons   Sensory neurons
Effector organ   Sensory receptors
Motor neuron

In a reflex arc, stimuli are detected by __(1)_, causing the production of action potentials that are carried to the central nervous system by _(2)_. Within the central nervous system, sensory neurons usually synapse with _(3)_. These neurons synapse with _(4)_, which carries action potentials to the _(5)_.

1. _____
2. _____
3. _____
4. _____
5. _____

B. Match these parts of a
reflex arc with the correct
parts labeled in Figure 8-4:

Association neuron
Effector organ
Motor neuron
Sensory neuron
Sensory receptor

1. _____

2. _____

3. _____

4. _____

5. _____

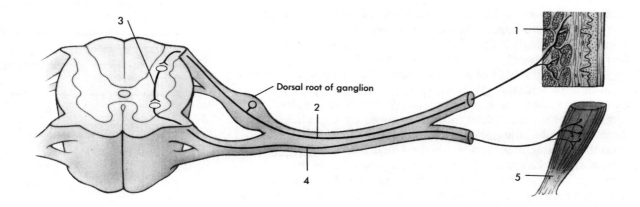

Dorsal root of ganglion

Meganges

"Three connective tissue layers, the meninges, surround and protect the brain and spinal cord.**"**

Match these terms with the
correct statement or definition:

Arachnoid layer
Dura mater
Dural sinus

Pia mater
Subarachnoid space

_____ 1. Most superficial and thickest of the meninges.

_____ 2. Spaces within the dura mater that collect blood from the small veins
of the brain.

_____ 3. Thin, wispy middle meningeal layer.

_____ 4. Meningeal layer that is very tightly bound to the surface of the brain
and spinal cord.

_____ 5. Space between the arachnoid layer and the pia mater which is
filled with cerebrospinal fluid.

Ventricles and Cerebrospinal Fluid

“_The CNS contains fluid-filled cavities called ventricles._**”**

Using the terms provided, complete the following statements:

Arachnoid granulations
Central canal
Cerebral aqueduct
Choroid plexus

Fourth ventricle
Hydrocephalus
Lateral ventricle
Third ventricle

Each cerebral hemisphere contains a relatively large cavity, the (1). The two lateral ventricles connect through small openings to a smaller (2) in the center of the diencephalon. A narrow canal, the (3) connects the third ventricle to the (4) located at the base of the cerebellum. The fourth ventricle is continuous with the (5) of the spinal cord and empties into the subarachnoid space. Cerebrospinal fluid (CSF), which bathes the brain and spinal cord is produced by a specialized membrane called the (6) in each of the ventricles. CSF passes from the subarachnoid space into the blood through the (7) in a dural sinus. Blockage of the cerebral aqueduct or of the openings between the fourth ventricle and the subarachnoid space can result in (8), a condition in which CSF accumulates and increases pressure on the brain.

1. _____
2. _____
3. _____
4. _____
5. _____
6. _____
7. _____
8. _____

Peripheral Nervous System

“_The peripheral nervous system consists of sensory and motor neurons._**”**

Match these terms with the correct statement or definition:

Cranial
Motor

Sensory
Spinal

_____

_____

_____

_____

1. Part of the peripheral nervous system with 12 pairs of nerves.

2. Part of the peripheral nervous system with 31 pairs of nerves.

3. Neurons that collect information and carry it to the CNS.

4. Neurons that convey information from the CNS to muscles and glands.

Cranial Nerves

"*There are three general categories of cranial nerve function: sensory, motor, and parasympathetic.*"

Match the cranial nerve with the correct function:

Olfactory (I)   Facial (VII)
Optic (II)   Vestibulocochlear (VIII)
Oculomotor (III)   Glossopharyngeal (IX)
Trochlear (IV)   Vagus (X)
Trigeminal(V)   Accessory (XI)
Abducens (VI)   Hypoglossal (XII)

_____ 1. Smell.

_____ 2. Controls four eye muscles, constricts the pupil, and thickens the lens.

_____ 3. Sensory to face and teeth.

_____ 4. Two nerves that are sensory to taste and parasympathetic to salivary glands.

_____

_____ 5. Hearing and balance.

_____ 6. Motor to tongue muscles.

_____ 7. Parasympathetic to viscera of thorax and abdomen.

_____ 8. Two nerves, each of which is motor to one eye muscle.

_____

Spinal Nerves

"*The spinal nerves arise along the spinal cord from the union of the dorsal root and ventral root.*"

Using the terms provided, complete the following statements:

Autonomic   Lumbosacral
Brachial   Phrenic
Cervical   Plexuses

All the spinal nerves are mixed nerves, containing both sensory and motor nerve fibers. However, most spinal nerves also have nerve fibers from the _(1)_ system. Most of the spinal nerves are organized into _(2)_, in which nerves come together and then separate. The _(3)_ plexus innervates the neck and posterior head, but one of the most important branches of this plexus is the _(4)_ nerve, which innervates the diaphragm. The _(5)_ plexus has branches that innervate the upper limb, whereas branches of the _(6)_ plexus innervate the lower limb.

1. _____

2. _____

3. _____

4. _____

5. _____

6. _____

Autonomic Nervous System

66_Motor neurons can be divided into two systems, the somatic and autonomic nervous systems._**99**

A. Match these terms with the correct statement or definition:

Autonomic system
Somatic system

_____ 1. One motor neuron extends from the CNS to skeletal muscle.

_____ 2. Two motor neurons extend to smooth muscle, cardiac muscle, and glands.

_____ 3. Usually unconsciously controlled.

B. Match these terms with the correct statement or definition:

Parasympathetic division
Sympathetic division

_____ 1. Increases heart and respiration rate, release of glucose, and inhibits digestive activities.

_____ 2. Stimulates vegetative activities such as digestion, defecation, and urination; slows heart and respiration rate.

_____ 3. Cell bodies located in brainstem nuclei of certain cranial nerves, or in S2 to S4 regions of spinal cord gray matter.

_____ 4. Nerve fibers extend to sympathetic chain ganglia, or to ganglia located closer to target organs.

_____ 5. Neurons from the ganglia extend a relatively short distance to the target organs.

_____ 6. All neurons secrete acetylcholine onto their target organ.

_____ 7. Most neurons secrete norepinephrine onto their target organ.

 Some people use biofeedback or meditation methods to relax by learning to reduce the heart rate or change the pattern of brain waves.

Nervous System Disorders

"Nervous system disorders can have widespread effects throughout the body."

A. Match these disorders with the correct description:

Alzheimers's disease   Multiple sclerosis (MS)
Cerebral palsy   Parkinson's disease
Cerebrovascular accident (CVA)   Senility
Myasthenia gravis   Tumor

_____ 1. Inflammation and immune response cause damage to myelin sheath.

_____ 2. The synapses between motor neurons and skeletal muscles are attacked by the immune system.

_____ 3. Defects in motor functions or coordination resulting from abnormal brain development or birth-related injury.

_____ 4. Caused by a lesion in the basal ganglia; characterized by muscular rigidity, tremor, and a slow, shuffling gait.

_____ 5. Destroys or compresses brain tissue.

_____ 6. A stroke; can be caused by a thrombosis (clot), embolism (moving clot), or vasospasm, resulting in an infarct.

_____ 7. General intellectual deficiency, mental deterioration, and memory loss resulting from several specific disease states.

_____ 8. Severe type of senility, often affecting people under age 50.

B. Match these disorders with the correct description:

Anesthesia   Neuralgia
Encephelitis   Neuritis
Epilepsy   Poliomyelitis
Herpes   Tetanus
Meningitis

_____ 1. Group of brain disorders with seizure episodes in common.

_____ 2. Severe spasms of throbbing or stabbing pain along a nerve pathway; common forms affect trigeminal, facial, or sciatic nerves.

_____ 3. Inflammation of a nerve; results from injury, infection, or other causes.

_____ 4. Loss of sensation; can be induced to facilitate medical treatment.

_____ 5. Inflammation of the brain, most often caused by a virus.

_____ 6. Inflammation of the meninges; may be caused by a virus or bacteria.

_____ 7. Toxin from bacteria prevents muscle relaxation; body becomes rigid.

_____ 8. Viral disease in which viruses reside in ganglia of sensory nerves; lesions produce cold sores, genital lesions, chicken pox, and shingles.

_____ 9. Viral infection; damages motor neurons extending to skeletal muscles.

1. Name three types of neurons according to function, and give their function.

2. List two types of cells found in the nervous system.

3. List three components of a synapse.

4. List the four major parts of the brain.

5   List four lobes of the cerebrum, and give an important function of each.

6. List the three meninges surrounding the brain and spinal cord.

7. List the three general categories of cranial nerve function.

8. Name the two divisions of the autonomic nervous system and list three differences between them.

Place the letter corresponding to the correct answer in the space provided.

_____ 1. The part of the nervous system that transmits impulses from the CNS to skeletal muscle is the
a. somatic nervous system.
b. autonomic nervous system.
c. central nervous system.
d. sensory division.

_____ 2. Neurons that are found primarily within the CNS, and conduct action potentials from one nerve cell to another nerve cell are
a. sensory neurons.
b. motor neurons
c. association neurons.

_____ 3. Which of the following statements about neuroglia is NOT correct?
a. Neuroglia are more numerous than neurons.
b. Neuroglia cells form myelin sheaths around axons.
c. Neuroglia cells hold neurons in place.
d. Neuroglia help to establish the blood-brain barrier.
e. Neuroglia release neurotransmitters.

_____ 4. Clusters of nerve cell bodies within the PNS are called
a. nuclei.
b. nodes of Ranvier.
c. myelin sheaths.
d. ganglia.

_____ 5. Myelin sheaths
a. increase the speed of action potentials.
b. form synapses.
c. prevent action potentials at the node of Ranvier.
d. cause action potentials to pass the entire length of the axon.

_____ 6. Concerning conditions in a resting cell membrane
a. there are more potassium ions outside the cell than inside.
b. there are more sodium ions inside the cell than outside.
c. the outside of the cell membrane is more positive than the inside.
d. the sodium-potassium exchange pump moves potassium ions out of the cell.
e. all of the above

_____ 7. To produce an action potential in a neuron,
a. depolarization must occur.
b. the threshold level of membrane potential must be reached.
c. the cell membrane must become permeable to sodium ions.
d. all of the above.

_____ 8. Neurotransmitters are released in a synapse and bind to
a. the axon ending.
b. the synaptic cleft.
c. vesicles.
d. receptors on the muscle fiber.

_____ 9. Important centers for control of heart rate, blood vessel diameter, breathing, swallowing and coughing are located in the
a. cerebrum.
b. cerebellum.
c. medulla oblongata.
d. diencephalon.

_____ 10. Our conscious state is maintained by activity generated in the
a. cerebellum.
b. reticular activating system.
c. cerebral cortex.
d. medulla.

_____11. The part of the brain connected to the pituitary gland is the
a. hypothalamus.
b. medulla oblongata.
c. pons.
d. cerebellum.
e. thalamus

_____12. The area in the parietal lobe receiving pain, touch, and temperature sensations is the
a. motor area.
b. auditory area.
c. general sensory area.
d. prefrontal area.

_____13. Which of the following is responsible for producing the muscle movements necessary for speech?
a. Broca's area
b. premotor area
c. auditory area
d. general sensory area

_____14. The largest connection between the right and left hemisphere of the cerebrum is the
a. basal ganglia.
b. limbic system.
c. corpus callosum.
d. cerebellum.

_____15. All sensory neurons entering the spinal cord
a. enter through the dorsal root.
b. have their cell bodies in the dorsal root ganglia.
c. are part of a spinal nerve.
d. all of the above

_____16. Given the following parts of the nervous system:
1. ascending nerve tract
2. asociation neuron in spinal cord gray matter
3. general sensory area
4. thalamus

Arrange the parts in the correct order that action potentials would pass through them after the finger is stuck with a pin.
a. 1,2,3,4
b. 1,2,4,3
c. 2,1,3,4
d. 2,1,4,3

_____17. Given the following parts of the nervous system:
1. descending nerve tract
2. motor areas
3. motor neuron in spinal cord gray matter
4. prefrontal area

Arrange the parts in the correct order that action potentials would pass through them to cause movement of skeletal muscle.
a. 2,4,1,3
b. 2,4,3,1
c. 4,2,1,3
d. 4,2,3,1

_____18. Given the following components of a reflex arc:
1. association neuron
2. effector organ
3. motor neuron
4. sensory neuron
5. sensory receptor

The correct sequence for these components in the operation of a reflex arc is
a. 1,4,5,3,2
b. 2,1,4,5,3
c. 4,5,1,2,3
d. 5,3,1,4,2
e. 5,4,1,3,2

_____19. The outermost meninges layer is a thick, tough membrane called the
a. dura mater.
b. arachnoid.
c. pia mater.
d. subarachnoid layer.

_____20. There are ____ pairs of cranial nerves, and ____ pairs of spinal nerves.
a. 12, 24
b. 31, 12
c. 12, 31
d. 10, 12

_____21. A collection of spinal nerves that join together and then separate after leaving the spinal cord is called a
a. ganglion.
b. nucleus.
c. projection nerve.
d. plexus.

_____22. Which of the following is expected if the sympathetic nervous system is stimulated?
a. Blood flow to the digestive organs increases.
b. Respiration and sweating increase.
c. Heart rate decreases.
d. Glucose release from the liver decreases.
e. both a and b

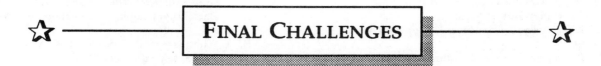

FINAL CHALLENGES

Use a separate sheet of paper to complete this section.

1. Given two series of neurons, explain why action potentials could be propagated along one series more rapidly than the other series.

2. Although alcohol has effects on other areas of the brain, it has a considerable effect on cerebellar function. What kinds of motor tests would help reveal the drunken condition?

3. Would a patient with Parkinson's disease be expected to have reduced or exaggerated reflexes? Explain.

4. A baby is born with enlarged lateral and third ventricles but a normal fourth ventricle. Describe the defect and its location.

5. Are the neuron cell bodies found in the following locations parasympathetic or sympathetic?
A. Cranial nerve nuclei
B. Lateral horn of thoracic spinal cord gray matter.
C. Lateral horn of sacral spinal cord gray matter.
D. Chain ganglia.
E. Ganglia in the wall of an organ.

6. Two patients are admitted to the hospital. According to their charts, both had herniated disks that were placing pressure on the roots of the sciatic nerve. One patient had pain in the sole of the foot; the other patient had pain on the dorsal surface of her foot. Explain how the same condition could produce such different symptoms.

ANSWERS TO CHAPTER 8

Divisions of the Nervous System
1. Central nervous system; 2. Peripheral nervous system; 3. Sensory division; 4. Somatic nervous system; 5. Autonomic nervous system

Cells of the Nervous System
A. 1. Cell body; 2. Dendrites; 3. Axons; 4. Nerve fibers
B. 1. Sensory neurons; 2. Motor neurons; 3. Association neurons
C. 1. Glioma; 2. Myelin sheaths; 3. Increase; 4. Nodes of Ranvier; 5. Blood-brain barrier
D. 1. Dendrite; 2. Axon; 3. Neuroglia cell; 4. Node of Ranvier; 5. Cell body

Organization of Nervous Tissue
1. Gray matter; 2. Cortex; 3. Nuclei; 4. Ganglion; 5. White matter; 6. Nerve tracts; 7. Nerves

Action Potentials
1. Positively charged; 2. Resting membrane potential; 3. Action potential; 4. Nerve impulse; 5. Myelinated; 6. Unmyelinated; 7. Node of Ranvier

Understanding Action Potentials
1. K^+ ions; 2. Na^+ ions; 3. Depolarization; 4. Repolarization; 5. Action potential; 6. Sodium-potassium exchange pump; 7. All-or-none; 8. Threshold

The Synapse
1. Synaptic cleft; 2. Neurotransmitters; 3. Acetylcholine and Norepinephrine; 4. Acetylcholinesterase

Central Nervous System
1. Brain; 2. Spinal cord; 3. Brain

Brainstem
1. Medulla oblongata; 2. Pons; 3. Midbrain; 4. Reticular activating system

Diencephalon
A. 1. Thalamus; 2. Hypothalamus; 3. Hypothalamus
B. 1. Hypothalamus; 2. Pituitary gland; 3. Thalamus

Cerebrum
A. 1 Gyri; 2. Sulci
B. 1. Frontal lobe; 2. Parietal lobe; 3. Occipital lobe; 4. Temporal lobe

Cerebral Cortex
1. General sensory area; 2. Motor area; 3. Broca's area; 4. Aphasia; 5. Prefrontal area

Memory
1. Short-term memory; 2. Long-term memory; 3. Memory engrams

Right and Left Cerebral Hemispheres
1. Right cerebral hemisphere; 2. Corpus callosum; 3. Left cerebral hemisphere

Basal Ganglia and Cerebellum
1. Basal ganglia; 2. Cerebellum; 3. Cerebellum

Spinal Cord
A. 1. Gray matter; 2. White matter; 3. Dorsal root; 4. Dorsal root ganglia; 5. Spinal nerve; 6. Ventral root
B. 1. White matter; 2. Dorsal root; 3. Dorsal root ganglion; 4. Ventral root; 5. Gray matter

Reflexes
A. 1. Sensory receptors; 2. Sensory neurons; 3. Association neurons; 4. Motor neurons; 5. Effector organ
B. 1. Sensory receptor; 2. Sensory neuron; 3. Association neuron; 4. Motor neuron; 5. Effector organ

Meninges
1. Dura mater; 2. Dural sinus; 3. Arachnoid layer; 4. Pia mater; 5. Subarachnoid space

Ventricles and Cerebrospinal Fluid
1. Lateral ventricle; 2. Third ventricle; 3. Cerebral aqueduct; 4. Fourth ventricle; 5. Central canal; 6. Choroid plexus; 7. Arachnoid granulations; 8. Hydrocephalus

Peripheral Nervous System
1. Cranial; 2. Spinal; 3. Sensory; 4. Motor

Cranial Nerves
1. Olfactory (I); 2. Oculomotor (III); 3. Trigeminal (V); 4. Facial (VII) and glossopharyngeal (IX); 5. Vestibulocochlear (VIII); 6. Hypoglossal (XII); 7. Vagus (X); 8. Trochlear (IV) and Abducens (VI)

Spinal Nerves
1. Autonomic; 2. Plexuses; 3. Cervical; 4. Phrenic; 5. Brachial; 6. Lumbosacral

Autonomic Nervous System
A. 1. Somatic system; 2. Autonomic system; 3. Autonomic system
B. 1. Sympathetic division; 2. Parasympathetic division; 3. Parasympathetic division; 4. Sympathetic division; 5. Parasympathetic division; 6. Parasympathetic division; 7. Sympathetic division

Nervous System Disorders
A. 1. Multiple sclerosis; 2. Myasthenia gravis; 3. Cerebral palsy; 4. Parkinson's disease; 5. Tumor; 6. Cerebrovascular accident (CVA); 7. Senility; 8. Alzheimer's disease
B. 1. Epilepsy; 2. Neuralgia; 3. Neuritis; 4. Anesthesia; 5. Encephelitis; 6. Meningitis; 7. Tetanus; 8. Herpes; 9. Poliomyelitis

1. Sensory neurons: conduct action potentials to CNS; Motor neurons: conduct action potentials away from CNS; Association neurons: conduct action potentials from one nerve cell to another nerve cell, primarily in the CNS.
2. Neurons and neuroglia
3. Axon ending of a neuron, synaptic cleft, and muscle fiber
4. Brainstem, diencephalon, cerebrum, and cerebellum
5. Frontal lobe: voluntary motor function, motivation, aggression, mood; parietal lobe: reception and evaluation of most sensory information such as pain, temperature, pressure (touch) and taste; occipital lobe: reception and integration of visual input; temporal lobe: evaluates olfactory and auditory input, plays an important role in memory, and in involved with abstract thought and judgement
6. Dura mater, arachnoid, and pia mater
7. Motor, sensory, and parasympathetic
8. Sympathetic division: prepares for physical activity, many neurons synapse in sympathetic chain ganglia, second neuron is longer than in parasympathetic, neurotransmitter for second neuron is usually norepinephrine, cell bodies for sympathetic neurons are between T1 and L2 in the spinal cord; parasympathetic division: prepares for vegetative functions, neurons synapse close to or on the effector organ, neurotransmitter is acetylcholine for all neurons, cell bodies for parasympathetic neurons are located in brainstem nuclei or S2 to S4 regions of the spinal cord.

MASTERY LEARNING ACTIVITY

1. A. The somatic nervous system transmits impulses to skeletal muscle. The autonomic nervous system regulates the activities of smooth muscle, cardiac muscle and glands. The central nervous system refers to the brain and spinal cord, and the sensory division transmits action potentials from the body to the CNS.

2. C. Association neurons are found primarily within the CNS, and conduct action potentials from one nerve cell to another nerve cell. Sensory neurons conduct action potentials away from the CNS, and motor neurons conduct action potentials to the CNS.

3. E. Neuroglia do not release neurotransmitters, only neurons do. All of the other statements about neuroglia are correct.

4. D. Clusters of nerve cell bodies in the PNS are called ganglia. Clusters of nerve cell bodies in the CNS are called nuclei. Nodes of Ranvier and myelin sheaths are related to neuroglia that are associated with the axons of neurons.

5. A. Myelin sheaths increase the speed of action potentials. In axons with myelin sheaths, action potentials only occur at the nodes of Ranvier; action potentials do not pass the entire length of the axon. Myelin sheaths are not part of synapses.

6. C. In a resting cell, the outside of the cell membrane is positive compared to the inside of the membrane. There are more sodium ions outside the cell than inside, and more potassium ions inside the cell than outside. The sodium-potassium exchange pump actively transports sodium ions out of the cell and potassium ions into the cell.

7. D. All of the conditions listed are necessary for an action potential to occur.

8. D. Neurotransmitters are released from the axon ending, diffuse across the synaptic cleft, and bind to receptors on the muscle fiber. Vesicles are the storage structures for the neurotransmitter before it is released.

9. C. The medulla oblongata has control centers for all the functions listed.

10. B. The reticular activating system is a loose network of nuclei scattered throughout the brainstem. This system has an important role in arousing and maintaining consciousness.

11. A. The hypothalamus is the location where the pituitary gland is attached to the brain.

12. C. Sensory input such as pain, touch, and temperature terminates in the general sensory area. The motor area is concerned with the control of skeletal muscles. The auditory area is the area in which the sensation of sound is perceived. The prefrontal area is the anterior portion of the frontal lobes and is involved in motivation and foresight to plan and initiate movements.

13. A. Broca's area initiates the series of movements necessary for speech. Then the prefrontal areas and auditory areas have functions as noted in question 12 above.

14. C. The corpus callosum is the largest of the connections (nerve tracts) joining the two cerebral hemispheres.

15. D. Sensory neurons enter through the dorsal root; the cell bodies of these neurons are in the dorsal root ganglia, and sensory and motor neurons join to form a spinal nerve.

16. D. An action potential, stimulated by a sensory neuron, first travels through an association neuron in spinal cord gray matter. The association neuron synapses with ascending nerve tracts, through which an action potential travels to the thalamus. From the thalamus, an action potential travels to a general sensory area in the cerebral cortex.

17. C. The motivation and foresight to plan movements occurs in the prefrontal area. From the prefrontal area, action potentials travel to the motor areas of the cerebral cortex, which are responsible for initiating action potentials for voluntary skeletal muscle movement. Action potentials then travel down descending nerve tracts, which synapse with motor neurons in the spinal cord gray matter. An action potentials travel through motor neurons to skeletal muscle.

18. E. The pathway an action potential would follow in a reflex is: sensory receptor, sensory neuron, association neuron, motor neuron, effector organ.

19. A. There are three layers of connective tissue called meninges, which cover the entire central nervous system. From outside to inside, they are the dura mater, arachnoid, and pia mater.

20. C. There are 12 pairs of cranial nerves and 31 pairs of spinal nerves.

21. D. A plexus is a collection of spinal nerves that join and then separate. Ganglia are collections of cell bodies in the peripheral nervous system. Nuclei are collections of nerve cell bodies in the central nervous system.

22. B. The sympathetic nervous system prepares the body for activity. Respiration and sweating increase, heart rate increases, and blood sugar levels increase. Meanwhile, processes not immediately necessary for activity are inhibited such as blood flow to digestive organs.

 FINAL CHALLENGES

1. If one series of neurons had more neurons, it would have more synapses, which should slow down the rate of action potential propagation. Also, if one series were unmyelinated and the other myelinated, the unmyelinated series would be slower.

2. Because the cerebellum acts to match intended movements with actual movements, reduced cerebellar function results in an inability to point precisely to an object (such as one's nose). It also results in poor balance.

3. Parkinson's disease is a basal ganglion disorder that results in decreased inhibition of muscles. Therefore, the neurons to muscles become overstimulated, resulting in muscular rigidity and tremors. The same overstimulated state also results in exaggerated reflexes.

4. The enlarged lateral and third ventricles suggest that the pressure being exerted by CSF is too high. Because the fourth ventricle is normal, it is logical to assume that a blockage in the cerebral aqueduct has prevented the outflow of CSF from the lateral and third ventricles to the fourth ventricle, where it can exit the brain. The condition is called internal hydrocephaly.

5. A. parasympathetic; B. sympathetic; C. parasympathetic;
D. sympathetic; E. parasympathetic

6. The sciatic nerve is composed of two nerves, the tibial, which innervates the skin of the sole of the foot, and the common peroneal, which innervates the skin of the dorsal foot. When different parts of the nerve are compressed by a herniated disk, pain is felt in different locations.

The General and Special Senses

FOCUS: The general senses are distributed throughout the body and include pain, temperature, touch, pressure, and proprioception (sense of position). Receptors for the general senses are found in the skin, tendons, ligaments, muscles, and body organs. The special senses involve localized organs in the head. These organs have very specialized sensory cells, and include smell (olfaction), taste, sight (vision), hearing, and balance (equilibrium).

CONTENT LEARNING ACTIVITY

Introduction

66 *The senses are the means by which the brain receives information about the outer world and the body.* **99**

Match these terms with the correct statement or definition:

Action potential
Central nervous system

General senses
Special senses

_____ 1. Produced by sensory receptors in response to a stimulus.

_____ 2. Site where sensations are actually perceived.

_____ 3. Category that includes pain, temperature, touch, pressure, and position awareness.

_____ 4. Category that includes taste, smell, sight, hearing, and balance.

_____ 5. Sensory receptors for this category of sensation are found throughout the body.

Understanding Pain

"Pain is a sensation characterized by a group of unpleasant perceptual and emotional experiences."

Match these terms with the correct statement or definition:

Diffuse and aching	Localized and sharp
Endorphins	Pain tolerance
Free nerve endings	Phantom pain
General anesthesia	Referred pain
Local anesthesia	Touch receptor

_____ 1. Type of receptor responsible for pain sensations.

_____ 2. Superficial pain is better localized than deep pain because superficial pain also activates this type of receptor.

_____ 3. Type of pain that is rapidly conducted from sensory receptors to the central nervous system.

_____ 4. Type of pain that is slowly conducted from sensory receptors to the central nervous system.

_____ 5. Type of anesthesia produced by injection of a drug near a nerve.

_____ 6. Type of anesthesia produced by drugs that suppress the reticular activating system.

_____ 7. Chemicals produced by the body that interfere with action potential transmission in nerve tracts.

_____ 8. Ability to endure pain; different from one person to the next.

_____ 9. Painful sensation in a region of the body that is not the source of the pain stimulus.

_____ 10. Perception of pain in a structure that is not present, such as an amputated finger.

Olfaction

"Olfaction occurs in response to airborne molecules that enter the nasal cavity."

Match these terms with the correct statement or definition:

Olfactory bulb	Olfactory nerves
Olfactory area	Olfactory neurons

_____ 1. Specialized cells in the epithelium of the nasal cavity; airborne molecules bind to receptors on these cells, producing action potentials.

_____ 2. Formed by axons from the olfactory neurons.

_____ 3. Receives the olfactory nerves.

_____ 4. Part of the temporal lobes that receives action potentials from the olfactory bulbs.

Taste

66*Taste occurs in response to molecules dissolved in fluid within the oral cavity.*99

Match these terms with the correct statement or definition:

Back of tongue   Taste area
Papillae   Taste bud
Side of tongue   Tip of tongue

_____ 1. Sensory structure that detects taste stimuli.

_____ 2. Enlargements on the surface of the tongue; contain taste buds.

_____ 3. Part of the parietal lobes that receives taste sensations conducted through cranial nerves.

_____ 4. Taste buds located on this part of the tongue respond most strongly to sweet and salt tastes.

_____ 5. Taste buds located on this part of the tongue respond most strongly to bitter tastes.

_____ 6. Taste buds located on this part of the tongue respond most strongly to sour tastes.

☞ Many taste sensations are strongly influenced by olfactory sensations.

Vision: Accessory Structures

66*Accessory structures protect, lubricate, and move the eye.*99

Match these terms with the correct statement or definition:

Conjunctiva   Extrinsic eye muscles
Eyebrows   Lacrimal gland
Eyelids   Nasolacrimal duct

_____ 1. Prevent perspiration from running down the forehead into the eyes.

_____ 2. Protect the eye from foreign objects and lubricate the eye by spreading tears.

_____ 3. Thin, mucous membrane that covers the inner surface of the eyelids and the anterior surface of the eye.

_____ 4. Produces tears.

_____ 5. Collects excess tears from the medial corner of the eye and empties them into the nasal cavity.

_____ 6. Move the eyeball.

☞ Strabismus is a condition in which one eye or both eyes are directed medially ("cross-eyed") or laterally ("cock-eyed")

Anatomy of the Eye

"The wall of the eye has three layers."

A. Match these terms with the correct statement or definition:

Cornea   Outer layer
Inner layer   Sclera
Middle layer

_____

_____

_____

_____

_____

1. Layer of the eye consisting of the sclera and cornea.

2. Layer of the eye consisting of choroid, ciliary body, and iris.

3. Layer of the eye consisting of the retina.

4. Firm, white, outer posterior five sixths of the eye; maintains the shape of the eye and provides an attachment site for the extrinsic eye muscles.

5. Transparent, anterior one sixth of the eye; allows light to enter the eye.

B. Match these terms with the correct statement or definition:

Choroid   Lens
Ciliary body   Pupil
Iris

_____

_____

_____

_____

_____

1. Middle layer of the eye that covers the posterior sclera; contains melanin that prevents light reflection.

2. Contains smooth muscles that attach to the lens by suspensory ligaments.

3. Flexible, biconvex, transparent disc.

4. Contains smooth muscles that regulate the amount of light entering the eye.

5. Opening in the iris through which light passes.

C. Match these terms with the correct statement or definition:

Cones
Rhodopsin
Rods

_____

_____

_____

1. Photoreceptor cells that are very sensitive to light and function in dim light; do not provide a very sharp image.

2. Photoreceptor cells responsible for color vision; provide very sharp images.

3. Photosensitive molecule in rods that changes shape when struck by light; vitamin A is necessary for its manufacture.

 There are three types of cones, each sensitive to a different color: blue, green, or red.

D. Match these terms with the correct statement or definition:

Aqueous humor   Visual area
Optic disc   Vitreous humor
Optic nerve

_____ 1. Blind spot of the eye; place where the optic nerves pass through the wall of the eye.

_____ 2. Fills the anterior compartment of the eye; maintains pressure within the eye and provides nutrients to the inner surface of the eye.

_____ 3. Fills the posterior compartment of the eye; helps to maintain pressure within the eye and holds the lens and retina in place.

_____ 4. Carries action potentials toward the brain.

_____ 5. Part of the occipital lobes in which action potentials from the eyes are interpreted as vision.

☞ Glaucoma is a build-up of pressure within the eye that can result in blindness.

E. Match these terms with the correct parts labeled in Figure 9-1:

Choroid   Optic nerve
Ciliary body   Pupil
Cornea   Retina
Iris   Sclera
Lens   Suspensory ligaments

1. _____

2. _____

3. _____

4. _____

5. _____

6. _____

7. _____

8. _____

9. _____

10. _____

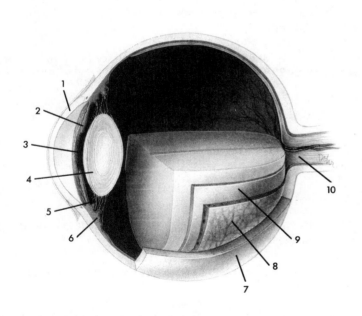

Functions of the Complete Eye

"The eye functions much like a camera."

A. Match these terms with the
   correct statement or definition:

Concave lens   Focusing
Convex lens   Refraction

_____ 1. Bending of light rays as they pass through a lens.

_____ 2. Type of lens that causes light rays to diverge.

_____ 3. Type of lens that causes light rays to converge.

_____ 4. Act of causing light rays to converge to form an image.

B. Match these terms with the
   correct statement or definition:

Accommodation   Lens
Contracted   Relaxed
Cornea

_____ 1. Two parts of the eye primarily responsible for focusing light.

_____

_____ 2. Part of the eye that accomplishes fine adjustments in focusing by
                              changing shape.

_____ 3. Process of allowing the lens to assume a more spherical (round) form;
                              enables the eye to focus objects that are closer than 20 feet.

_____ 4. Condition of the smooth muscles of the ciliary body during
                              accommodation.

_____ 5. Condition of the smooth muscles of the ciliary body for distant vision
                              (greater than 20 feet).

Eye Disorders

"Knowledge of eye disorders is clinically important."

A. Match these terms with the correct statement or definition:

Astigmatism   Neonatal gonorrheal ophthalmia
Chlamydial conjunctivitis Presbyopia
Hyperopia   Trachoma
Myopia

_____ 1. Two infections acquired by newborns passing through the birth canal.

_____

_____ 2. Leading cause of blindness in the world; causes scarring of the cornea.

_____ 3. Ability to see close objects but not far objects; nearsightedness; corrected with a concave lens.

_____ 4. Ability to see far away objects but not close-up objects; farsightedness; corrected with a convex lens.

_____ 5. Decreased ability to accommodate.

_____ 6. Image is not clearly focused because the cornea or lens is not uniformly curved.

B. Match these terms with the correct statement or definition:

Cataracts   Diabetes
Color blindness Glaucoma
Detached retina

_____ 1. Clouding of the lens.

_____ 2. Excessive pressure in the eye resulting from a buildup of aqueous humor.

_____ 3. This disorder can cause blindness because of the blood vessel degeneration and hemorrhage in the retina.

_____ 4. Separation of the photoreceptor layer of the retina from the layer of pigmented cells in the retina; results in impaired vision or blindness.

_____ 5. Absence of perception of one or more colors because cone cells do not function properly.

Auditory Structures and Their Functions

The external, middle, and inner ear are involved in hearing; balance is a function of the inner ear.

A. Match these terms with the correct statement or definition:

Auricle (pinna)   External auditory meatus
Cerumen   Tympanic membrane

_____ 1. Fleshy part of the external ear on the outside of the head.

_____ 2. Passageway that leads to the tympanic membrane.

_____ 3. Earwax, which helps to prevent foreign objects from reaching the tympanic membrane.

_____ 4. Thin membrane that separates the external and middle ear; vibrates in response to sound waves; also called the eardrum.

B. Match these terms with the correct statement or definition:

Auditory ossicles   Oval window
Auditory tube   Round window
Mastoid air cells

_____ 1. Opening between the middle and inner ear; the auditory ossicles (stapes) attaches to the opening.

_____ 2. Membrane-covered opening between the middle and inner ear with nothing attached to the membrane.

_____ 3. Spaces in the mastoid process that are connected to the middle ear.

_____ 4. Structure that enables air pressure to be equalized between the outside air and the middle ear; the Eustachian tube.

_____ 5. Ear bones that transmit vibrations of the tympanic membrane to the oval window; the malleus, incus and stapes.

C. Match these terms with the correct statement or definition:

Cochlea
Semicircular canals
Vestibule

_____ 1. Part of the inner ear involved with hearing.

_____ 2. Two parts of the inner ear involved with balance.

_____

D. Match these terms with the correct parts labeled in Figure 9-2:

Auditory ossicles
Auditory tube
Auricle
Cochlea
External auditory meatus
External ear
Inner ear
Middle ear
Oval window
Round window
Semicircular canals
Tympanic membrane
Vestibule
Vestibulocochlear nerve

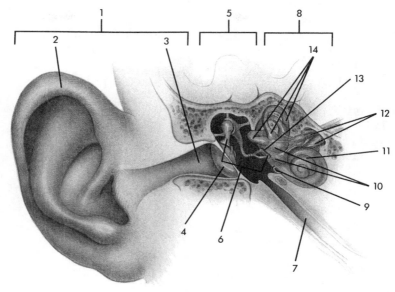

1. _____   6. _____   11. _____

2. _____   7. _____   12. _____

3. _____   8. _____   13. _____

4. _____   9. _____   14. _____

5. _____   10. _____

Hearing

66*Sound waves are converted into action potentials that the brain interprets as sound.*99

A. Using the terms provided, complete the following statements:

Action potentials
Auditory ossicles
Auricle
Hair cells
Organ of Corti

Oval window
Round window
Tympanic membrane
Vestibulocochlear nerve

1. _____

2. _____

3. _____

4. _____

5. _____

6. _____

7. _____

8. _____

9. _____

Sound waves in the air are collected by the (1) and conducted by the external auditory meatus to the (2), which vibrates. The vibrations are transferred from the tympanic membrane by the (3) to an opening into the inner ear, the (4). Consequently, vibrations are produced in the fluid of the inner ear. These vibrations cause slight movement of the (5), which results in the bending and stretching of (6). The movement of the hair cells results in the production of (7) that are conducted to the brain by the (8). In the temporal lobe the action potentials are interpreted as sound in the auditory area. The vibrations produced in the fluid of the inner ear are eventually absorbed by the membrane of the (9).

B. Match these terms with
the correct parts labeled
in Figure 9-3:

Auditory ossicles
Auditory tube
Canals
Organ of Corti
Oval window
Round window
Tympanic membrane
Vestibulocochlear nerve

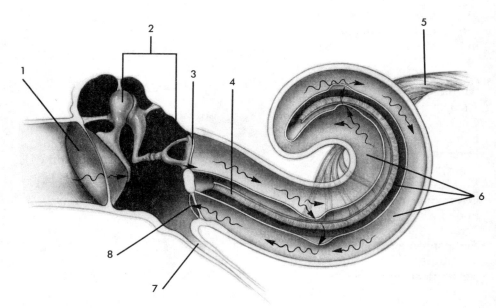

1. _____   4. _____   7. _____

2. _____   5. _____   8. _____

3. _____   6. _____

Equilibrium

❝*The sense of equilibrium has two components: static and kinetic equilibrium.*❞

Match these terms with the
correct statement or definition:

Crista ampullaris   Maculae
Kinetic equilibrium   Static equilibrium

_____ 1. Type of equilibrium associated with the vestibule and involves
evaluating the position of the head relative to the surface of the
earth.

_____ 2. Type of equilibrium associated with the semicircular canals and
involves evaluating the change in rate of head movements.

_____ 3. Specialized epithelium of the vestibule; consists of hair cells
embedded in a gelatinous mass.

_____ 4. Specialized epithelium of the semicircular canals; consists of hair
cells embedded in a gelatinous mass.

Ear Disorders

Knowledge of ear disorders is clinically important.

Match these terms with the correct statement or definition:

Conduction deafness
Otitis media
Motion sickness

Otosclerosis
Sensorineural deafness

_____ 1. Deafness resulting from a mechanical difficulty in the transmission of sound waves from the outer ear to the organ of Corti.

_____ 2. Deafness resulting from damage to the organ of Corti or nerve pathways.

_____ 3. Type of deafness usually helped the most by hearing aids.

_____ 4. Bone growth over the oval window or auditory ossicles; results in conduction deafness.

_____ 5. Infection of the middle ear.

_____ 6. Nausea and weakness caused by stimulation of the semicircular canals.

QUICK RECALL

1. List the sensations produced through the general senses and the special senses.

2. List the four basic tastes detected by the taste buds.

3. List the structures found in each layer of the eye.

4. Name the two types of photoreceptor cells in the retina, and state two important functional differences between them.

5. Name the two compartments of the eye and the substances found in each compartment.

6. Name the two structures of the ear that are closed off by a membrane.

7. List the three sensory structures of the inner ear and state their function.

MASTERY LEARNING ACTIVITY

Place the letter corresponding to the correct answer in the space provided.

_____ 1. Which of the following is an example of a general sense?
   a. balance
   b. hearing
   c. pain
   d. vision

_____ 2. Pain sensations
   a. are conducted rapidly or slowly.
   b. can be blocked by endorphins.
   c. can be felt at body locations that are not a source of pain stimulation.
   d. all of the above

_____ 3. Olfactory neurons
   a. have axons that combine to form the olfactory nerves.
   b. connect to the olfactory bulb.
   c. have receptors that react with molecules dissolved in fluid.
   d. all of the above

_____ 4. Taste
   a. is detected by receptors called papillae.
   b. sensations from the tongue are conducted to the brain by spinal nerves.
   c. can be divided into four basic types: sour, salty, bitter, and sweet.
   d. all of the above

_____ 5. Tears
   a. are released onto the surface of the eye near the medial corner of the eye.
   b. in excess are removed by the Eustachian tube.
   c. are carried to the back of the throat and swallowed.
   d. lubricate and clean the eye, and protect against eye infections.

_____ 6. Given the following structures:
1. choroid
2. retina
3. sclera

Choose the arrangement that list the structures in the order a pin would pass through them going from the outside of the eye to the inside.
a. 1, 2, 3
b. 2, 3, 1
c. 3, 1, 2
d. 3, 2, 1

_____ 7. Which of the following structures is correctly matched with it function?
a. choroid - contains rods and cones responsible for vision
b. ciliary body - changes the size of the pupil
c. cornea - attachment site for the extrinsic eye muscles
d. iris - regulates the amount of light entering the eye

_____ 8. As light intensity decreases,
a. the diameter of the pupil decreases and rod activity decreases.
b. the diameter of the pupil decreases and rod activity increases.
c. the diameter of the pupil increases and rod activity increases.
d. the diameter of the pupil increases and rod activity decreases.

_____ 9. Light striking this structure is the first step in the process that allows us to see color.
a. cone
b. optic disk
c. melanin in the choroid
d. rod

_____ 10. Given the following structures:
1. lens
2. aqueous humor
3. vitreous humor
4. cornea

Choose the arrangement that lists the structures in the order that light entering the eye would encounter them.
a. 1, 2, 3, 4
b. 1, 4, 2, 3
c. 4, 1, 2, 3
d. 4, 2, 1, 3
e. 4, 3, 1, 2

_____ 11. Given the following events:
1. light strikes a rod
2. production of action potential
3. rhodopsin changes shape

Choose the arrangement that list the events in the order they occur
a. 1, 2, 3
b. 1, 3, 2
c. 2, 1, 3
d. 2, 3, 1

_____ 12. Aqueous humor
a. is the pigment responsible for the black color of the choroid.
b. is produced by the iris.
c. returns to the blood through the nasolacrimal duct.
d. in excess can cause cataracts.
e. produces pressure that keeps the eye inflated.

_____ 13. Assume that you are looking at an object that is 30 feet away from you. If you suddenly look at an object that is 1 foot away, which of the following events would occur?
a. smooth muscles in the ciliary body contract and the lens flattens
b. smooth muscles in the ciliary body contract and the lens becomes more spherical (rounder)
c. smooth muscles in the ciliary body relax and the lens flattens
d. smooth muscles in the ciliary body relax and the lens becomes more spherical

_____ 14. Which structure is found within or is part of the external ear?
a. auditory ossicles
b. auditory tube
c. auricle
d. mastoid air cells

_____ 15. Which structure enables air pressure to be equalized between the outside air and the middle ear cavity?
a. auditory ossicles
b. auditory tube
c. auricle
d. tympanic membrane

_____16. Which of the following is correctly
matched with its function?
a. cochlea - kinetic balance
b. semicircular canals - static balance
c. tympanic membrane - vibrates in
response to sound waves
d. vestibule - hearing

_____17. Given the following structures:
1. auditory ossicles
2. auricle
3. external auditory meatus
4. tympanic membrane

Choose the arrangement that lists the
structures in the order sound waves
coming from the outside would encounter
them.
a. 2, 3, 1, 4
b. 2, 3, 4, 1
c. 3, 2, 1, 4
d. 3, 4, 2, 1

_____18. Given the following structures:
1. fluid of the inner ear
2. organ of Corti
3. oval window
4. round window

Choose the arrangement that lists the
membranes in the order sound coming
from the outside would encounter them.
a. 3, 1, 2, 4
b. 3, 2, 1, 4
c. 4, 1, 2, 3
d. 4, 2, 1, 3

_____19. Which of the following structures have
hair cells?
a. crista ampullaris
b. maculae
c. organ of Corti
d. all of the above

_____20. Damage to the sensory structures in the
semicircular canals would
a. damage the macula.
b. damage the organ of Corti.
c. affect the ability to detect the
position of the head relative to the
ground.
d. affect the ability to detect changes
in the rate of head movements.

FINAL CHALLENGES

Use a separate sheet of paper to complete this section.

1. Describe all the sensations involved in biting
into an apple. Which of those sensations are
general and which are special?

2. A man has gallstones that cause distention and
cramping of his gallbladder. What kind of
pain would he experience and where would the
pain be located? Explain.

3. An anatomy and physiology student conducted a
test on the sense of taste. A volunteer was
blindfolded and then asked to identify by taste
items placed on her tongue. For each of the
items, predict the likelihood that the
volunteer will correctly identify the item.
a. sugar water placed on the tip of the tongue
b. unsweetened tea placed on the tip of the
tongue
c. the tongue is dried and a few sugar crystals
are placed on the tip of the tongue

4. The main way that people "catch" colds is
through their hands. After touching an object
contaminated with the cold virus, the person
transfers the virus to the nasal cavity where it
causes an infection. Other than the obvious
entry of the virus through the nose, how could
the virus get into the nasal cavity?

5. At the battle of Bunker Hill, it could have been
said "Don't shoot until you see the _____ of
their eyes."

6. Compared to their normal position when a
person is standing upright, if the hair cells in
the macula were stretched (not bent), in what
position would a person be relative to the
ground? Explain.

ANSWERS TO CHAPTER 9

Introduction
1. Action potential; 2. Central nervous system;
3. General senses; 4. Special senses; 5. General senses

Understanding Pain
1. Free nerve endings; 2. Touch receptor;
3. Localized and sharp; 4. Diffuse and aching;
5. Local anesthesia; 6. General anesthesia;
7. Endorphins; 8. Pain tolerance; 9. Referred pain;
10. Phantom pain

Olfaction
1. Olfactory neurons; 2. Olfactory nerves; 3. Olfactory bulb; 4. Olfactory area

Taste
1. Taste bud; 2. Papillae; 3. Taste area; 4. Tip of tongue; 5. Back of tongue; 6. Side of tongue

Vision: Accessory Structures
1. Eyebrows; 2. Eyelids; 3. Conjunctiva; 4. Lacrimal gland; 5. Nasolacrimal duct; 6. Extrinsic eye muscles

Anatomy of the Eye
A. 1. Outer layer. 2. Middle layer; 3. Inner layer;
4. Sclera; 5. Cornea
B. 1. Choroid; 2. Ciliary body; 3. Lens; 4. Iris; 5. Pupil
C. 1. Rods; 2. Cones; 3. Rhodopsin
D. 1. Optic disc; 2. Aqueous humor; 3. Vitreous humor; 4. Optic nerve; 5. Visual area
E. 1. Cornea; 2. Iris; 3. Pupil; 4. Lens; 5. Suspensory ligaments; 6. Ciliary body; 7. Sclera; 8. Choroid;
9. Retina; 10. Optic nerve

Functions of the Complete Eye
A. 1. Refraction; 2. Concave lens; 3. Convex lens;
4. Focusing
B. 1. Cornea and lens; 2. Lens; 3. Accommodation;
4. Contracted; 5. Relaxed

Eye Disorders
A. 1. Neonatal gonorrheal ophthalmia and chlamydial conjunctivitis; 2. Trachoma;
3. Myopia; 4. Hyperopia; 5. Presbyopia;
6. Astigmatism
B. 1. Cataracts; 2. Glaucoma; 3. Diabetes;
4. Detached retina; 5. Color blindness

Auditory Structures and Their Functions
A. 1. Auricle (pinna); 2. External auditory meatus;
3. Cerumen; 4. Tympanic membrane
B. 1. Oval window; 2. Round window; 3. Mastoid air cells; 4. Auditory tube; 5. Auditory ossicles
C. 1. Cochlea; 2. Semicircular canals and vestibule
D. 1. External ear; 2. Auricle; 3. External auditory meatus; 4. Tympanic membrane; 5. Middle ear;
6. Auditory ossicles; 7. Auditory tube; 8. Inner ear;
9. Round window; 10. Vestibule; 11. Cochlea;
12. Vestibulocochlear nerve; 13. Oval window;
14. Semicircular canals

Hearing
A. 1. Auricle; 2. Tympanic membrane; 3. Auditory ossicles; 4. Oval window; 5. Organ of Corti; 6. Hair cells; 7. Action potentials; 8. Vestibulocochlear nerve; 9. Round window
B. 1. Tympanic membrane; 2. Auditory ossicles;
3. Oval window; 4. Organ of Corti;
5. Vestibulocochlear nerve; 6. Canals; 7. Auditory tube; 8. Round window

Equilibrium
1. Static equilibrium; 2. Kinetic equilibrium;
3. Maculae; 4. Crista ampullaris

Ear Disorders
1. Conduction deafness; 2. Sensorineural deafness;
3. Conduction deafness; 4. Otosclerosis; 5. Otitis media; 6. Motion sickness

1. General senses: pain, temperature, touch, pressure, and proprioception (sense of position); Special senses: smell, taste, sight, hearing, and balance (equilibrium)
2. Sour, salty, bitter, and sweet.
3. Outer layer: sclera and cornea; Middle layer: choroid, ciliary body, and iris; Inner layer: retina
4. Rods: very sensitive to light (function in dim light) but do not provide sharp images; Cones: less sensitive to light (require more light to operate than rods), provide sharp images, responsible for color vision
5. Anterior compartment: aqueous humor: Posterior compartment: vitreous humor
6. External auditory meatus (tympanic membrane) and the round window
7. Cochlea: hearing; Vestibule: static balance; Semicircular canals: kinetic balance

1. C. General senses include pain, temperature, touch, pressure, and proprioception (sense of body position). Special senses include smell, taste, sight, hearing, and balance.

2. D. Localized, sharp pain is transmitted rapidly, whereas diffuse, aching, burning pain is transmitted slowly. Endorphins are chemicals that block pain transmission within nerve tracts of the central nervous system. Referred pain is pain felt at a location that is not the source of the pain stimulus.

3. D. Olfactory neurons react to molecules dissolved in fluid to produce action potentials. The action potentials travel through the olfactory nerves, which are formed from the axons of the olfactory neurons, to the olfactory bulbs.

4. C. There are four basic tastes: sour, salty, bitter, and sweet. Taste buds on papillae detect taste. Taste sensations (action potentials) are carried from the tongue to the brain through cranial nerves.

5. D. Tears lubricate and clean the eye, and protect against eye infections. Tears are released from the lacrimal gland located in the superior, lateral corner of the eye and are collected in the medial corner by the nasolacrimal duct and conducted to the nasal cavity. The Eustachian tube, or auditory tube, connects from the middle ear to the throat.

6. C. The correct order is sclera, choroid, retina.

7. D. The iris contains smooth muscles that regulate the amount of light entering the eye by changing the diameter of the pupil. The retina contains rods and cones, the ciliary body changes the shape of the lens, and the sclera is the attachment site for the extrinsic eye muscles.

8. C. As light intensity decreases, the pupil dilates (increases in diameter), and the rods become more active. Both of these changes increase the ability to see when light intensity is low.

9. A. Cones are the photoreceptor cells responsible for color vision, whereas rods are the photoreceptor cells responsible for vision in dim light. The optic disc consists of axons, does not respond to light, and is called the blind spot. Melanin in the choroid absorbs light so that light is not reflected and does not interfere with vision.

10. D. Light passes through the cornea, aqueous humor, lens, and vitreous humor.

11. B. Light strikes a rod, changing the shape of rhodopsin, resulting in the production of an action potential.

12. E. Aqueous humor keeps the eye inflated. It is produced by the ciliary body and returns to the blood through a vein that encircles the cornea. Excess aqueous humor causes glaucoma, an elevated pressure within the eye. The choroid is black because of the pigment melanin. Cataracts are a clouding of the lens.

13. B. These are the events of accommodation. The smooth muscles in the ciliary body contract and the lens becomes more spherical when we look at objects that are closer than 20 feet to us.

14. C. The auricle is part of the external ear; the auditory tube and auditory ossicles are part of the middle ear; the mastoid air cells are spaces within the mastoid process that are connected to the middle ear.

15. B. The auditory (Eustachian) tube connects the middle ear to the pharynx (throat). Air movement through the auditory tube equalizes pressure between the middle ear and outside air.

16. C. The tympanic membrane vibrates in response to sound waves. The cochlea is involved with hearing, the vestibule with static balance, and the semicircular canals with kinetic balance.

17. B. The auricle collects sound waves that travel down the external auditory meatus to the tympanic membrane, causing it to vibrate. The auditory ossicles, which attach to the tympanic membrane then vibrate.

18. A. The stapes attaches to the oval window. Vibrations of the other auditory ossicles cause movement of the stapes, which produces vibrations in the fluid of the inner ear. As a result there is movement of the organ of Corti. Eventually the vibrations reach the round window, which absorbs or dampens the vibrations.

19. D. All of the sensory structures of the inner ear function as a result of the bending or stretching of hair cells.

20. D. The cristae ampullaris in the semicircular canals are responsible for detecting changes in the rate of head movements. The maculae in the vestibule detect the position of the head relative to the ground. The organ of Corti is responsible for hearing.

 FINAL CHALLENGES

1. General sensations: touch, pressure, temperature, and awareness of the mandible (jaw) changing position; special sensations: taste, smell, vision, and hearing (crunch of the apple bite).

2. Pain from visceral organs is not well localized because of the absence of touch receptors. Pain from visceral organs is normally perceived as a diffuse pain. The pain from the gallbladder is often referred to the right, anterior, superior surface of the abdomen (right upper quadrant). This referred pain can be quite intense.

3. The tip of the tongue has taste buds that react strongly to sugar and salt, so it is likely that the volunteer will correctly identify the sugar solution. The back of the tongue is better at identifying bitter tastes such as unsweetened tea. The volunteer probably will not correctly identify the tea. Because taste buds are stimulated by dissolved substances, drying the tongue should prevent correct identification of the sugar crystals.

4. When a contaminated hand rubs the eyes, the virus can be introduced into tears on the conjunctiva. From there the virus can spread into the lacrimal canaliculi, and pass through the nasolacrimal duct into the nasal cavity.

5. The actual quote is "Don't shoot until you see the whites of their eyes." The sclerae form the whites of the eyes.

6. If the hair cells of the maculae are stretched (not bent), a person is upside down. Gravity acts on the gelatinous mass in the macula and causes the hair cells to stretch. If the hair cells are bent, it indicates that the person is lying on his side.

The Endocrine System

FOCUS: The endocrine system is one of the major regulatory systems in the body, along with the nervous system. However, the endocrine system responds more slowly, and has a longer-lasting, more general effect on the body than the nervous system. Endocrine glands are ductless glands that release hormones into the blood, where they are carried to target tissues and either increase or decrease cell activity. Each hormone influences only those cells that have receptors for that hormone. The secretion of hormones is controlled by negative-feedback mechanisms. The major endocrine glands are the pituitary, thyroid, parathyroids, adrenals, pancreas, testes, ovaries, thymus, and pineal body.

CONTENT LEARNING ACTIVITY

The Endocrine System

"*The endocrine system is made up of all the body's endocrine glands.***"**

Match these terms with the correct statement or definition:

Endocrine glands Receptors
Exocrine glands Target tissues
Hormones

_____ 1. Glands without ducts that secrete chemicals into the blood, e.g., thyroid gland and adrenal glands.

_____ 2. Glands that secrete their products into ducts, e.g., sweat glands and salivary glands.

_____ 3. Chemicals that are secreted by endocrine glands into the blood and act on target tissues at another site in the body.

_____ 4. Cells that have receptors for, and respond to, a specific hormone.

_____ 5. Location on cells where hormones can bind.

The Pituitary and Hypothalamus

"The pituitary gland is a pea-sized gland located below the hypothalamus."

A. Match these terms with the correct statement or definition:

Anterior pituitary   Pituitary (hypophysis)
Hypothalamus   Posterior pituitary
Infundibulum

_____ 1. Small gland that rests in a depression of the sphenoid bone; controls the function of many other glands in the body.

_____ 2. Stalk that connects the pituitary gland to the hypothalamus.

_____ 3. Part of the pituitary controlled by releasing hormones produced by the hypothalamus.

_____ 4. Part of the pituitary controlled by nerve cells that originate in the hypothalamus.

_____ 5. Part of the brain through which emotions or stress can influence the endocrine system.

B. Match these terms with the correct statement or definition:

Nerve cells in hypothalamus Releasing hormones
Hypothalamic-pituitary portal
 system

_____ 1. Chemicals produced by the hypothalamus that control secretion of hormones from the anterior pituitary.

_____ 2. Capillary beds and veins that transport releasing hormones to the anterior pituitary.

_____ 3. Source of hormones released from axon nerve endings in the posterior pituitary.

Hormones of the Anterior Pituitary

66*The anterior pituitary secretes several hormones that affect other glands.*99

Match these hormones
with the correct function
or description:

Adrenocorticotropic hormone (ACTH)
Follicle-stimulating hormone (FSH)
Growth hormone (GH)
Luteinizing hormone (LH)
Melanocyte-stimulating hormone (MSH)
Prolactin (PRL)
Thyroid-stimulating hormone (TSH)

_____ 1. Stimulates the growth of bones, muscles, and other organs by increasing protein synthesis; favors fat breakdown.

_____ 2. Increases the secretion of cortisol from the adrenal glands; increases skin pigmentation.

_____ 3. Causes ovulation in females and sex hormone secretion in males and females.

_____ 4. Stimulates development of oocytes in the ovary and sperm cells in the testis.

_____ 5. Promotes development of the breast during pregnancy and causes production of milk after pregnancy.

☞ In a young person, too little growth hormone produces a pituitary dwarf, and too much growth hormone produces giantism. Too much growth hormone after bone growth in length is complete produces acromegaly.

Hormones of the Posterior Pituitary

66*Posterior pituitary hormones are produced in nerve cell bodies in the hypothalamus, and released*99 *from their axon endings in the posterior pituitary.*

Match these hormones
with the correct function
or description:

Antidiuretic hormone (ADH)
Oxytocin

_____ 1. Increases water retention by the kidneys and constriction of blood vessels; also called vasopressin.

_____ 2. Causes contraction of muscles of the uterus and milk "let-down."

The Thyroid Gland

66 *The thyroid gland is made up of two lobes of thyroid tissue connected by a narrow band.* *99*

A. Match these terms with the correct statement or definition:

Calcitonin   Thyroid follicles
Parafollicular cells   Thyroid hormones

_____ 1. Small spheres of cuboidal epithelium that synthesize and store thyroid hormones.

_____ 2. Hormones produced in the thyroid gland that regulate the rate of metabolism in the body.

_____ 3. Scattered cells located between thyroid follicles.

_____ 4. Hormone synthesized by parafollicular cells that influences calcium ion levels in the body.

B. Match these conditions with the correct symptom or condition:

Hyperthyroidism
Hypothyroidism

_____ 1. Cretinism in infants.

_____ 2. In adults, a reduced metabolic rate, sluggishness, a reduced ability to perform routine tasks, and myxedema.

_____ 3. Elevated metabolic rate, extreme nervousness, chronic fatigue, and exophthalmos.

_____ 4. Iodine deficiency and goiter.

C. Complete each statement by providing the missing word:

Decreases
Increases

_____ 1. Increased thyroid hormone production _____ TSH production.

_____ 2. Decreased TSH production _____ thyroid hormone production.

_____ 3. Excess TSH _____ the size of the thyroid gland.

_____ 4. Increased calcium level in blood _____ calcitonin secretion.

_____ 5. Calcitonin _____ calcium level in the blood.

The Parathyroid Glands

"Four tiny parathyroid glands are embedded in the posterior wall of the thyroid gland."

A. Complete each statement by providing the missing word:

Decreases
Increases

_____ 1. PTH _____ the breakdown of bone tissue to release calcium into the blood.

_____ 2. PTH _____ the rate at which calcium is lost in the urine.

_____ 3. PTH _____ active vitamin D formation.

_____ 4. Active vitamin D _____ absorption of calcium by the small intestine.

_____ 5. Decreased calcium level in the blood _____ PTH production.

B. Match these conditions with the correct description:

Muscular weakness   Tetany
Rickets

_____ 1. Condition resulting from a chronic lack of vitamin D; occurs most often in young people.

_____ 2. Condition in which blood levels of calcium decrease below their normal range of values and spontaneous action potentials occur.

_____ 3. Condition in which blood levels of calcium increase above their normal range of values; cell membranes have decreased permeability to sodium ions.

☞ PTH is more important than calcitonin in regulating blood calcium levels.

The Adrenal Glands

"The adrenal glands are two small glands, each located on top of a kidney."

Match these terms with the correct statement or definition:

Adrenal cortex
Adrenal medulla

_____ 1. Inner part of the adrenal gland, which releases epinephrine and norepinephrine.

_____ 2. Outer part of the adrenal gland, which releases steroid hormones.

The Adrenal Medulla

"Responses to hormones from the adrenal medulla reinforce the effect of the" sympathetic nervous system.

Complete each statement by providing the missing word:

Decrease
Increase

_____ 1. Adrenal medulla hormones (epinephrine and norepinephrine) cause a(n) _____ in blood flow to internal organs.

_____ 2. Adrenal medulla hormones _____ heart rate and blood pressure.

_____ 3. Adrenal medulla hormones _____ the metabolic rate in skeletal muscle, cardiac muscle, and nervous tissue.

_____ 4. Adrenal medulla hormones _____ the size of airway passages.

_____ 5. Adrenal medulla hormones _____ the release of glucose and fatty acids into the blood.

☞ Adrenal medulla hormones are referred to as the fight-or-flight hormones because they prepare the body for vigorous physical activity.

The Adrenal Cortex

"Three classes of steroid hormones are secreted by the adrenal cortex."

A. Match these terms with the correct statement or definition:

Aldosterone   Cortisone
Androgens   Glucocorticoids
Cortisol   Mineralocorticoids

_____ 1. Class of steroid hormones that help to regulate blood nutrient levels in the body.

_____ 2. Major glucocorticoid hormone.

_____ 3. Steroid often given as a medication to reduce inflammation.

_____ 4. Class of steroid hormones that help regulate blood volume and levels of potassium and sodium ions.

_____ 5. Major mineralocorticoid hormone.

_____ 6. Class of steroid hormones that stimulate the development of male sexual characteristics.

B. Complete each statement by providing the missing word:

Decreases
Increases

_____ 1. Cortisol _____ the breakdown of proteins and fat.

_____ 2. Cortisol _____ the use of amino acids and fatty acids as energy sources.

_____ 3. Stress _____ the secretion of cortisol.

_____ 4. Cortisol _____ the inflammatory response.

_____ 5. If ACTH increases, the secretion of cortisol _____.

C. Complete each statement by providing the missing word:

Decreases
Increases

_____ 1. Aldosterone _____ sodium ion and water retention by the body.

_____ 2. Aldosterone _____ potassium ion retention by the body.

_____ 3. Aldosterone secretion _____ when blood potassium levels increase.

_____ 4. Aldosterone secretion _____ when blood pressure or sodium levels decrease.

D. Match these conditions with the correct description:

Addison's disease
Cushing's syndrome

_____ 1. Results from abnormally low levels of aldosterone and cortisol; symptoms include low blood glucose, low blood pressure, and a large volume of urine produced.

_____ 2. Results from an excess secretion of aldosterone and cortisol; symptoms include high blood glucose, destruction of muscle, high blood pressure, and production of a small volume of urine.

The Pancreas

66 *Pancreatic islets secrete two hormones, insulin and glucagon, that help regulate blood levels of glucose.* 99

A. Match these terms with the correct statement or definition:

Glucagon   Pancreatic islets
Insulin   Satiety center

_____ 1. Endocrine cell clusters dispersed among exocrine cells in the pancreas.

_____ 2. Hormone secreted by pancreatic islets that decreases blood glucose levels.

_____ 3. Hormone secreted by pancreatic islets that increases blood glucose levels.

_____ 4. Area of the hypothalamus that regulates appetite.

B. Complete each statement by providing the missing word:

Decreases
Increases

_____ 1. If blood glucose level decreases below normal, the ability of the nervous system to function _____.

_____ 2. When blood glucose level decreases below normal, the breakdown of fat _____.

_____ 3. Increased breakdown of fat _____ the pH of the body fluids, leading to acidosis.

_____ 4. If blood glucose levels are too high, the volume of urine produced _____, resulting in dehydration.

C. Complete each statement by providing the missing word:

Decrease(s)
Increase(s)

_____ 1. Insulin causes glucose and amino acid uptake, glycogen synthesis, and fat synthesis in the body to _____.

_____ 2. Insulin causes blood glucose level to _____.

_____ 3. Glucagon _____ the breakdown of glycogen to glucose.

_____ 4. Glucagon causes blood glucose level to _____.

_____ 5. When blood glucose levels increase, insulin secretion _____.

_____ 6. When blood glucose levels decrease, glucagon secretion _____.

Understanding Diabetes Mellitus and Insulin Shock

"A lack of insulin secretion results in a condition called diabetes mellitus."

A. Complete each statement by providing the missing word:

Decrease(s)
Increase(s)

_____ 1. In people who have diabetes mellitus, the glucose uptake into tissues _____.

_____ 2. In people who have diabetes mellitus, blood glucose level _____.

_____ 3. In people who have diabetes mellitus, glucose is not available for metabolism, so breakdown of fats and proteins _____.

_____ 4. In people who have diabetes mellitus, appetite and thirst _____.

_____ 5. In people who have diabetes mellitus, energy level and amount of body tissue _____.

147

B. Match these terms with the correct statement or definition:

Insulin shock   Type II diabetes mellitus
Type I diabetes mellitus

_____ 1. Juvenile diabetes mellitus; caused by a lack of insulin secretion.

_____ 2. Maturity onset diabetes mellitus; results from the reduced ability of tissues to respond to insulin; condition produces obesity.

_____ 3. Malfunction of the brain that occurs when insulin levels are too high; may result in disorientation, convulsions, and loss of consciousness.

☞ Diabetes mellitus results from secretion of too little insulin by the pancreas, insufficient numbers of insulin receptors on target cells, or defective receptors that do not respond to insulin.

The Testes and Ovaries

❝The testes of the male and the ovaries of the female secrete sex hormones in addition to producing sperm cells or oocytes.**❞**

Using the terms provided, complete the following statements:

Decrease   Menstrual cycle
Estrogen and progesterone   Ovary
Increase   Testosterone

The main hormone produced by the testes in the male is (1) . It is responsible for growth and development of the male reproductive structures, a(n) (2) in muscle size and body hair, voice changes, and sex drive. In the female, (3) contribute to development and function of female reproductive structures and other female sexual characteristics. The female (4) is controlled by cyclic release of estrogens and progesterone from the (5) .

1. _____
2. _____
3. _____
4. _____
5. _____

Other Hormones

❝Several other glands and tissues produce hormones that have widespread effects in the body.**❞**

Match these hormones with the correct description:

Digestive hormones   Prostaglandins
Melatonin   Thymosin

_____ 1. Hormones produced in the lining of the stomach and small intestine that increase production of digestive juices and movement of food through the digestive tract.

_____ 2. Hormone produced in widespread tissues throughout the body that causes relaxation or contraction of smooth muscle, blood vessel dilation, swelling, and pain.

_____ 3. Assists in the development of white blood cells called T cells.

_____ 4. Pineal body hormone that decreases secretion of FSH and LH; linked to the onset of puberty.

Correctly match these endocrine glands with the hormone each secretes:

Adrenal cortex
Adrenal medulla
Anterior pituitary
Ovaries
Pancreas
Parathyroid glands

Pineal body
Posterior pituitary
Thymus gland
Thyroid gland (follicle cells)
Thyroid gland (parafollicular cells)
Testes

_____ 1. ACTH.

_____ 2. ADH.

_____ 3. Adrenal androgens.

_____ 4. Aldosterone.

_____ 5. Calcitonin.

_____ 6. Cortisol.

_____ 7. Epinephrine

_____ 8. Estrogen

_____ 9. FSH and LH

_____ 10. GH

_____ 11. Glucagon

_____ 12. Insulin

_____ 13. Melatonin

_____ 14. Oxytocin

_____ 15. Progesterone

_____ 16. Prolactin

_____ 17. PTH

_____ 18. Thymosin

_____ 19. Thyroxine

_____ 20. Testosterone

Place the letter corresponding to the correct answer in the space provided.

_____ 1. An endocrine gland
   a. lacks a duct.
   b. secretes chemicals to an internal or external surface of the body.
   c. produces sweat or saliva.
   d. all of the above.

_____ 2. Hormones
   a. can be proteins or lipids.
   b. can either increase or decrease cell activity.
   c. are distributed throughout the body by the blood.
   d. bind to receptors on target tissues.
   e. all of the above

_____ 3. The secretion of hormones
   a. is controlled by negative-feedback mechanisms.
   b. has rapid, short-term effects throughout the body.
   c. causes large increases or decreases in hormone levels in the blood.
   d. both b and c

_____ 4. The pituitary gland
   a. is connected to the brain by the infundibulum.
   b. controls the function of many other glands.
   c. has two parts.
   d. secretes hormones that affect growth and kidney function
   e. all of the above

_____ 5. Secretion of hormones from the anterior pituitary is controlled by
   a. releasing hormones produced by the hypothalamus.
   b. releasing hormones produced in the posterior pituitary.
   c. action potentials from the hypothalamus.
   d. action potentials from the posterior pituitary.

_____ 6. Hormones secreted in the posterior pituitary
   a. are produced in the thalamus.
   b. are transported to the posterior pituitary by the hypothalamic-pituitary portal system.
   c. include ADH and oxytocin.
   d. all of the above

_____ 7. Growth hormone
   a. increases the breakdown of fat.
   b. decreases protein synthesis.
   c. increases protein breakdown.
   d. all of the above

_____ 8. Excess growth hormone
   a. results in giantism if it occurs in children.
   b. causes acromegaly in adults.
   c. causes goiter.
   d. both a and b

_____ 9. LH and FSH
   a. are produced in the hypothalamus.
   b. production is increased by TSH.
   c. regulate growth and function of the gonads.
   d. inhibit the production of prolactin.

_____ 10. Which of the following would result from a thyroidectomy (removal of the thyroid gland)?
   a. increased calcitonin secretion
   b. increased TSH secretion
   c. increased thyroid hormone secretion
   d. increased GH secretion

_____ 11. If parathyroid hormone levels increase, which of the following would be expected?
   a. Breakdown of bone is increased.
   b. Calcium absorption from the small intestine is decreased.
   c. Calcium reabsorption from urine is decreased.
   d. Less active vitamin D would be formed.

12. If a condition produced excessive secretion of the adrenal medulla, which of the following symptoms would you expect?
    a. low blood pressure
    b. decreased heart rate
    c. increased blood flow to internal organs
    d. increased glucose and fatty acids in the blood
    e. all of the above

13. A hormone secreted from the adrenal cortex is
    a. aldosterone.
    b. cortisol.
    c. androgen.
    d. a and b
    e. all of the above

14. Cortisol
    a. increases the breakdown of fats.
    b. increases the breakdown of proteins.
    c. increases blood sugar levels.
    d. decreases inflammation.
    e. all of the above

15. Aldosterone
    a. causes increased sodium retention in the body.
    b. causes increased water retention in the body.
    c. causes increased potassium retention in the body.
    d. a and b
    e. all of the above

16. In regard to regulation of aldosterone secretion, the adrenal cortex is most sensitive to
    a. blood sodium levels.
    b. blood potassium levels.
    c. blood calcium levels.
    d. blood cortisol levels

17. Insulin
    a. increases the uptake of glucose by target cells.
    b. increase uptake of amino acids by target cells.
    c. increases glycogen synthesis in liver and skeletal muscle cells.
    d. all of the above

18. If a person who has diabetes mellitus forgets to take an insulin injection, symptoms that may soon appear include
    a. acidosis.
    b. high blood glucose.
    c. increased urine production.
    d. increased thirst.
    e. all of the above

✮ ── FINAL CHALLENGES ── ✮

Use a separate sheet of paper to complete this section.

1. A young boy (6 years old) exhibited marked and rapid development of sexual characteristics. On examination his testicles were not found to be larger than normal, but his blood testosterone levels were elevated. As a mental exercise, a student nurse decided that she would propose a cure. She considered the symptoms and decided on surgery to remove an adrenal tumor. Explain why you agree or disagree with her diagnosis.

2. During pregnancy and/or lactation, parathyroid hormone may be elevated. Explain why this might occur and give a reason why it might be harmful.

3. A patient has pheochromocytoma, a condition in which a benign tumor causes excessive secretion of the adrenal medulla. Would you expect the pupils of the patient to be dilated or constricted? (Hint: pupil diameter is controlled by smooth muscle)

4. Given that a tumor often causes overproduction of a hormone, explain how a tumor in the pituitary gland could produce goiter.

5. Normally ACTH stimulates the adrenal cortex to secrete cortisol, and through a negative-feedback mechanism, cortisol inhibits ACTH secretion. In Addison's disease, there is a lower than normal secretion of cortisol. Would you expect the skin of a person with Addison's disease to have increased or decreased pigmentation? Explain.

ANSWERS TO CHAPTER 10

CONTENT LEARNING ACTIVITY

The Endocrine System
1. Endocrine glands; 2. Exocrine glands; 3. Hormones; 4. Target tissues; 5. Receptors

The Pituitary and Hypothalamus
A. 1. Pituitary (hypophysis); 2. Infundibulum; 3. Anterior pituitary; 4. Posterior pituitary; 5. Hypothalamus
B. 1. Releasing hormones; 2. Hypothalamic-pituitary portal system; 3. Nerve cells in hypothalamus

Hormones of the Anterior Pituitary
1. Growth hormone (GH), 2. Adrenocorticotropic hormone (ACTH), 3. Luteinizing hormone (LH), 4. Follicle-stimulating hormone (FSH), 5. Prolactin (PRL)

Hormones of the Posterior Pituitary
1. Antidiuretic hormone (ADH), 2. Oxytocin

The Thyroid Gland
A. 1. Thyroid follicles; 2. Thyroid hormones; 3. Parafollicular cells; 4. Calcitonin
B. 1. Hypothyroidism; 2. Hypothyroidism; 3. Hyperthyroidism; 4. Hypothyroidism
C. 1. Decreases; 2. Decreases; 3. Increases; 4. Increases; 5. Decreases

The Parathyroid Glands
A. 1. Increases; 2. Decreases; 3. Increases; 4. Increases; 5. Increases
B. 1. Rickets; 2. Tetany; 3. Muscular weakness

The Adrenal Glands
1. Adrenal medulla; 2. Adrenal cortex

The Adrenal Medulla
1. Decrease; 2. Increase; 3. Increase; 4. Increase; 5. Increase

The Adrenal Cortex
A. 1. Glucocorticoids; 2. Cortisol; 3. Cortisone; 4. Mineralocorticoids; 5. Aldosterone; 6. Androgens
B. 1. Increases; 2. Increases; 3. Increases; 4. Decreases; 5. Increases
C. 1. Increases; 2. Decreases; 3. Increases; 4. Increases
D. 1. Addison's disease; 2. Cushing's syndrome

The Pancreas
A. 1. Pancreatic islets; 2. Insulin; 3. Glucagon; 4. Satiety center
B. 1. Decreases; 2. Increases; 3. Decreases; 4. Increases
C. 1. Increase; 2. Decrease; 3. Increases; 4. Increase; 5. Increases; 6. Increases
D. 1. Increase; 2. Decrease; 3. Increases; 4. Increase; 5. Increases; 6. Increases; 7. Decrease

Understanding Diabetes Mellitus and Insulin Shock
A. 1. Decreases; 2. Increases; 3. Increases; 4. Increase; 5. Decreases
B. 1. Type I diabetes mellitus; 2. Type II diabetes mellitus; 3. Insulin shock

The Testes and Ovaries
A. 1. Testosterone; 2. Increase; 3. Estrogen and progesterone; 4. Menstrual cycle; 5. Ovary

Other Hormones
1. Digestive hormones; 2. Prostaglandins; 3. Thymosin; 4. Melatonin

1. Anterior pituitary
2. Posterior pituitary
3. Adrenal cortex
4. Adrenal cortex
5. Thyroid gland (parafollicular cells)
6. Adrenal cortex
7. Adrenal medulla
8. Ovaries
9. Anterior pituitary
10. Anterior pituitary

11. Pancreas
12. Pancreas
13. Pineal body
14. Posterior pituitary
15. Ovaries
16. Anterior pituitary
17. Parathyroid glands
18. Thymus
19. Thyroid gland (follicular cells)
20. Testes

MASTERY LEARNING ACTIVITY

1. A   An endocrine gland lacks a duct, releases its secretions into the blood, and produces hormones.

2. E.   All of the statements about hormones are correct.

3. A   The secretion of hormones is controlled by negative-feedback mechanisms, which prevent large increases or decreases of hormones in the blood. This keeps the body functioning within a narrow range of values consistent with maintaining homeostasis. Hormones are usually have slower, long-term effects compared to the nervous system.

4. E.   All of the statements about the pituitary gland are correct.

5. A   Releasing hormones from the hypothalamus travel through blood vessels in the hypothalamic-pituitary portal system and control the secretions of the anterior pituitary.

6. C.   Posterior pituitary hormones include ADH and oxytocin. Hormones secreted in the posterior pituitary are produced in the hypothalamus in neuron cell bodies, travel to the posterior pituitary within the axons of these neurons, and are released from the axons.

7. A   Growth hormone stimulates the growth of bones, muscles, and other organs by increasing protein synthesis. It also resists protein breakdown, and favors fat breakdown.

8. D.   Excess secretion of growth hormone in children can result in giantism. In adults, however, only certain bones respond to the growth hormone, resulting in a condition called acromegaly. Diabetes mellitus is caused by a lack of the effects of insulin.

9. C.   LH and FSH are hormones produced in the anterior pituitary that regulate growth and function of the gonads. They stimulate sex hormone production (LH), and production of sperm cells or oocytes (FSH).

10. B.   Removal of the thyroid gland would eliminate thyroid hormone production. In response to decreased thyroid hormone secretion, TSH is released in larger amounts. Calcitonin is released by the parafollicular cells of the thyroid, and removal of the thyroid stops calcitonin production.

11. A   Parathyroid hormone has effects that increase the calcium level in the blood. PTH increases the breakdown of bone, releasing more calcium into the blood, increases calcium reabsorption from the small intestine and urine, and increases active vitamin D formation.

12. D.   The adrenal medulla produces epinephrine and norepinephrine, which increase heart rate, blood pressure, blood flow to skeletal muscle, metabolic rate, and also increase the amounts of glucose and fatty acids in the blood. Blood flow to internal organs is decreased.

13. E.   The adrenal cortex secretes three major classes of steroids: glucocorticoids, of which cortisol is the major example, mineralocorticoids, of which aldosterone is the major example, and androgens.

14. E.   Cortisol increases the breakdown of fats and proteins and increase blood sugar levels. Cortisol also decreases the intensity of the inflammatory response.

15. D.   Aldosterone causes increased sodium and water retention, but eliminates potassium from the body.

16. B. Blood sodium and blood potassium levels both influence aldosterone secretion, but apparently the adrenal cortex is much more sensitive to changes in blood potassium levels. Blood calcium levels or blood cortisol levels do not influence aldosterone secretion.

17. D. Insulin promotes the uptake of glucose and amino acids, both of which can be used as an energy source. In the liver and in skeletal muscle, the glucose is stored as glycogen. In adipose tissue glucose is converted to fat. The amino acids can be used to synthesize proteins or glucose.

18. E. Symptoms consistent with insufficient insulin include acidosis, high blood glucose, increased urine production, thirst, and hunger.

 ☆ **FINAL CHALLENGES** ☆

1. Rapid sexual development in a prepubertal boy is indicative of excessive secretion of androgens. The two most likely tissues are the testes and adrenal glands. Since the testes are of normal size, it would be possible that removal of an adrenal tumor would cure the boy.

2. During pregnancy the fetus withdraws calcium from the mother's blood and during lactation calcium is lost with the milk. In both cases, lower blood calcium levels stimulate parathyroid hormone release. Consequently, calcium is released from the mother's bones, or is absorbed more efficiently from her small intestine and urine, helping to maintain blood calcium levels. The removal of calcium from her bones could result in weakened bones.

3. Pheochromocytoma results in overproduction of epinephrine and norepinephrine by the adrenal medulla. These chemicals produce the same effects as sympathetic nervous system stimulation (see Chapter 8). Consequently, one would expect dilated pupils.

4. A pituitary tumor can produce excess TSH. TSH production would not be influenced by the negative-feedback effect usually exerted by increased thyroid hormone. As a result, the excessive TSH would cause thyroid tissue to enlarge, producing a goiter.

5. You would expect increased skin pigmentation. Cortisol has a negative feedback effect on ACTH; therefore decreased cortisol production results in increased ACTH secretion. ACTH binds to melanocytes in the skin and increases skin pigmentation.

Blood

FOCUS: Blood consists of plasma and formed elements. The plasma is 92% water with dissolved or suspended molecules, including albumin, globulins, and clotting proteins. The formed elements include red blood cells, white blood cells, and platelets. The red blood cells transport oxygen and carbon dioxide, whereas the white blood cells protect the body against microorganisms and remove dead cells and debris from the body. Platelets prevent bleeding by forming platelet plugs and by producing factors involved in clotting.

CONTENT LEARNING ACTIVITY

Introduction

❝*Blood travels to and from the tissues and plays an important role in homeostasis.***❞**

Using the terms provided, complete the following statements:

Body temperature	Microorganisms
Carbon dioxide	Oxygen
Clotting	pH

Blood has many functions in the body. It transports _(1)_, nutrients, enzymes, and hormones to tissues, and carries _(2)_ and waste products away. Blood helps to regulate _(3)_ and _(4)_. It protects the body against _(5)_, and prevents blood loss by _(6)_.

1. _____

2. _____

3. _____

4. _____

5. _____

6. _____

 The total blood volume in the average adult is approximately 4 to 5 liters in females and 5 to 6 liters in males. Blood makes up approximately 8% of the body's total weight.

Plasma

"The liquid part of blood is called plasma."

Match these terms with the correct statement or definition:

Albumin   Plasma
Clotting proteins   Serum
Globulins   Water

_____ 1. Major component of plasma.

_____ 2. Helps to maintain the normal movement of water between blood and tissues.

_____ 3. Molecules that function in immunity.

_____ 4. Fluid formed when the proteins that produce clots are removed from plasma.

Formed Elements

"The formed elements are cells or parts of cells."

Match these terms with the correct statement or definition:

Platelets
Red blood cells
White blood cells

_____ 1. Responsible for 95% of the volume of the formed elements; also called erythrocytes.

_____ 2. Only formed element with a nucleus; also called leukocytes.

_____ 3. Cell fragments; also called thrombocytes.

Red Blood Cells

"Red blood cells live for about 120 days in males and 110 days in females."

A. Match these terms with the correct statement or definition:

Erythropoietin   Hemoglobin
Folic acid   Globin
Heme   Vitamin B_{12}

_____ 1. Molecule in red blood cells that transports oxygen.

_____ 2. Part of hemoglobin that contains iron to which oxygen binds.

_____ 3. Two substances required for the cell divisions necessary to produce red blood cells.

_____

_____ 4. When blood oxygen levels decrease the kidneys release this substance, which stimulates red blood cell production.

B. **M**atch these terms with the
   correct statement or definition:

Bilirubin   Jaundice
Erythropoietin   Macrophages
Iron

_____ 1. Cells in the liver and spleen that remove red blood cells from the blood.

_____ 2. Component of heme that is recycled.

_____ 3. Heme is converted into this chemical, which is excreted in bile.

_____ 4. Yellowish color of the skin resulting from a build-up of bilirubin.

_____ 5. Converted by intestinal bacteria to pigments that are responsible for the brown color of feces.

C. **M**atch these terms with
   the correct parts labeled
   in Figure 11-1:

Bilirubin   Hemoglobin
Globin   Iron
Heme

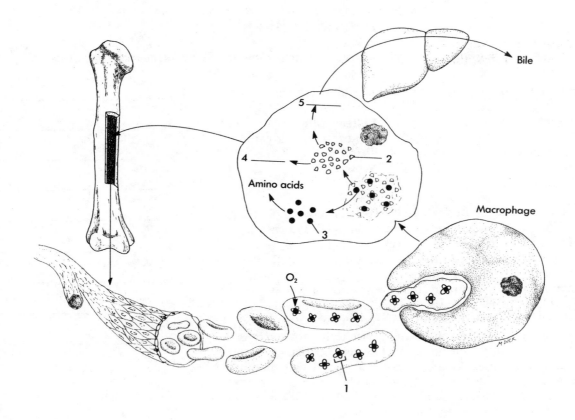

1. _____   3. _____   5. _____

2. _____   4. _____

157

White Blood Cells and Platelets

66 *White blood cells are nucleated cells that lack hemoglobin.* 99

Match these terms with the
correct statement or definition:

Basophil   Monocyte
Eosinophil   Neutrophil
Lymphocyte   Platelet

_____   1. White blood cell that phagocytizes microorganisms; forms pus when
                                they accumulate and die.

_____   2. White blood cell that becomes a macrophage.

_____   3. White blood cells involved in immunity; produce antibodies.

_____   4. Two white blood cells that release chemicals involved with
                                inflammation

_____

_____   5. Cell fragments that play a role in preventing blood loss.

Preventing Blood Loss

66 *Platelet plug and clot formation are very important to the maintenance of homeostasis.* 99

A. Using the terms provided, complete the following statements:

Calcium   Platelet plug
Clot   Prothrombin activator
Clotting factors   Thrombin
Fibrin   Vitamin K
Liver

Exposure of collagen in damaged tissue can activate platelets,
which form a (1) that can seal small tears in blood vessels.
Activated platelets also release (2) , prostaglandins, and
other chemicals necessary for blood clotting. Normally the
blood contains inactive proteins called (3) . These proteins
are activated by exposure to connective tissue or chemicals
released from injured tissues. Activation results in a series of
chemical reactions that lead to the production of (4) . This
substance converts prothrombin to (5) which in turn converts
fibrinogen into (6) . This network of protein fibers traps blood
cells, platelets, and fluid, and is called a (7) . This structure
can prevent blood loss from large tears in blood vessels. Many
of the clotting factors are produce by the (8) , and many of the
clotting factors require (9) for their synthesis.

1. _____

2. _____

3. _____

4. _____

5. _____

6. _____

7. _____

8. _____

9. _____

B. Match these terms with the correct statement or definition:

Anticoagulant
Aspirin
Embolus
Streptokinase
Plasmin
Thrombus
Tissue plasminogen activator

_____ 1. General term for a substance that prevents clots from forming.

_____ 2. Substance that prevents clots from forming by inhibiting prostaglandin synthesis by platelets.

_____ 3. Clot that forms within a blood vessel.

_____ 4. Detached clot or substance that floats through the circulatory system and becomes lodged in a blood vessel.

_____ 5. Formed from plasminogen, this substance breaks down fibrin.

_____ 6. Two substances used to break down a thrombus that caused a heart attack.

_____

Blood Grouping

"Red blood cells can be grouped according to the types of molecules they have on the outside of their cell membranes."

A. Match these terms with the correct statement or definition:

Agglutination
Antibodies
Antigens
Blood groups
Hemolysis
Transfusion

_____ 1. Transfer of blood into the blood of a patient.

_____ 2. Molecules on the surface of red blood cells that can bind to antibodies.

_____ 3. Molecules in the plasma that can combine with antigens on red blood cells; activate mechanisms that destroy the red blood cells.

_____ 4. Clumping together of red blood cells caused by antibodies combining with antigens.

_____ 5. Rupture of red blood cells.

_____ 6. Classes of red blood cells based on their surface antigens.

B. Match these terms with the correct statement or definition:

Donor   Type A blood
No reaction   Type B blood
Recipient   Type AB blood
Transfusion reaction   Type O blood

_____ 1. Type of blood that has A antigens and B antibodies.

_____ 2. Type of blood that does not have A or B antigens, but does have A and B antibodies.

_____ 3. Person who receives blood.

_____ 4. People with this type of blood have been called universal donors.

_____ 5. Result of giving a transfusion of type A blood to a person with type A blood.

_____ 6. Result of giving a transfusion of type A blood to a person with type B blood.

Understanding Rh Blood Groups

66 *The Rh blood group is so named because it was first studied in rhesus monkeys.* **99**

Match these terms with the correct statement or definition:

Anti-$Rh_o(D)$ Rh antibodies
Bilirubin   Rh-negative
Hemolytic disease of the newborn   Rh-positive

_____ 1. Type of blood that has certain Rh antigens on the surface of the red blood cells.

_____ 2. Disorder that results in agglutination and hemolysis of fetal blood because of different types of Rh blood in the mother and fetus.

_____ 3. Substance produced by the mother that promotes the agglutination and hemolysis reactions of HDN.

_____ 4. Type of blood the fetus has in HDN.

_____ 5. Type of blood the mother has in HDN.

_____ 6. Given to prevent HDN.

_____ 7. Infants with HDN are treated with fluorescent lights to break down this substance.

_____ 8. Type of Rh blood used for transfusion treatment of infants with HDN.

Diagnostic Blood Tests

" *Blood tests can prevent transfusion reactions and provide useful information about a patient's health.* **"**

Match these terms with the correct statement or definition:

Blood chemistry
Complete blood count
Hematocrit
Hemoglobin
Platelet count

Prothrombin time
Red blood cell count
Type and cross match
White blood cell count
White blood cell differential

_____ 1. Test used to prevent transfusion reactions.

_____ 2. Includes a red blood cell count, hemoglobin and hematocrit measurements, and white blood cell count.

_____ 3. This test would detect polycythemia.

_____ 4. Measures the volume of the red blood cells.

_____ 5. This test detects leukemia.

_____ 6. Test that determines the percentages of each of the five kinds of white blood cells.

_____ 7. This test detects thrombocytopenia.

_____ 8. Measures how long it takes for blood to start clotting.

_____ 9. Determines the composition of materials in plasma, such as glucose.

Some Disorders of the Blood

" *Anemia is a deficiency of normal hemoglobin in the blood.* **"**

A. Match these terms with the correct statement or definition:

Aplastic anemia
Folic acid deficiency anemia
Hemolytic anemia
Hemorrhagic anemia

Iron deficiency anemia
Pernicious anemia
Sickle cell anemia
Thalassemia

_____ 1. Inability of the red bone marrow to produce red blood cells.

_____ 2. Results from a lack of iron.

_____ 3. Results from inadequate vitamin B_{12}, usually caused by inadequate absorption of the vitamin because of inadequate intrinsic factor.

_____ 4. Anemia caused by loss of blood such as ulcers or menstrual bleeding.

_____ 5. Disorder in which red blood cells are destroyed at an excessive rate; HDN is an example.

_____ 6. Hereditary disorder resulting in inadequate hemoglobin production.

_____ 7. Hereditary disorder in which red blood cells assume an abnormal shape and rupture.

B. Match these terms with the correct statement or definition:

AIDS   Malaria
Hemophilia   Mononucleosis
Hepatitis   Septicemia
Leukemia

_____ 1. Genetic disorder in which clotting is abnormal or absent.

_____ 2. Type of cancer in which there is abnormal production of one more types of white blood cells.

_____ 3. Multiplication of microorganisms in the blood, sometimes called blood poisoning.

_____ 4. Rupture of red blood cells caused by a protozoan that is transmitted to humans by mosquito bites.

_____ 5. Viral disease that affects the salivary glands and lymphocytes; produces fever, sore throat, and swollen lymph nodes.

_____ 6. Blood is routinely tested for the viruses that cause these two diseases.

_____

QUICK RECALL

1. List six functions of blood.

2. Explain how a hemoglobin molecule transport oxygen.

3. Describe the fate of each part of the hemoglobin molecule when hemoglobin is broken down.

4. List the events that lead to increased red blood cell production when blood oxygen levels decrease.

5. List the five types of white blood cells and give a function of each.

6. Give two ways that platelets prevent blood loss.

7. Starting with the production of prothrombin activator, list the chemicals that result in the formation of a clot.

8. Describe the basic mechanisms responsible for transfusion reactions.

MASTERY LEARNING ACTIVITY

Place the letter corresponding to the correct answer in the space provided.

_____ 1. Which of the following is a function of blood?
   a. prevents fluid loss
   b. transports hormones
   c. carries oxygen to cells
   d. involved in the regulation of body temperature
   e. all of the above

_____ 2. Which of the following is NOT a component of plasma?
   a. albumin
   b. globulin
   c. fibrinogen
   d. platelets
   e. glucose

_____ 3. Red blood cells
   a. are the least numerous formed element in blood.
   b. are cylindrical shaped cells.
   c. are produced in yellow marrow.
   d. do not have a nucleus

_____ 4. Hemoglobin
   a. is responsible for the red color of red blood cells.
   b. is composed of heme and globin molecules.
   c. contains iron, which binds to and transports 97% of the oxygen in blood.
   d. all of the above

_____ 5. Which of the components of hemoglobin is correctly matched with its fate following the destruction of a red blood cell?
   a. heme - reused to form new hemoglobin molecule
   b. globin - broken down into amino acids
   c. iron - mostly secreted in bile
   d. all of the above

_____ 6. Erythropoietin
   a. is produced mainly by the liver.
   b. inhibits red blood cell production.
   c. production increases when blood oxygen levels decreases.
   d. increases iron absorption from the small intestine.

_____ 7. The most common type of white blood cell, which functions to phagocytize microorganisms and other substances?
   a. lymphocyte
   b. macrophage
   c. neutrophil
   d. basophils

_____ 8. Monocytes
   a. are the smallest white blood cells.
   b. give rise to macrophages.
   c. typically accumulate to form pus at sites of infection.
   d. all of the above

_____ 9. Given the following chemicals:
   1. fibrin
   2. prothrombin activator
   3. thrombin

   Choose the arrangement that lists the chemicals in the order they are formed during clot formation.
   a. 1, 2, 3
   b. 1, 3, 2
   c. 2, 1, 3
   d. 2, 3, 1

_____ 10. The chemical that is involved in the breakdown of a clot?
   a. fibrinogen
   b. anticoagulants
   c. clotting factors
   d. plasmin

_____ 11. Type B blood
   a. has type A antigens.
   b. has type B antibodies.
   c. causes a transfusion reaction when given to a person with type A blood.
   d. all of the above

_____ 12. Rh-negative mothers are given anti-Rh_O(D) immune globulin (RhoGam) injections in order to
   a. initiate the synthesis of Rh antibodies in the mother.
   b. initiate the synthesis of Rh antibodies in the fetus.
   c. prevent the mother from producing Rh antibodies.
   d. prevent the fetus from producing Rh antibodies.

_____ 13. The diagnostic blood test that would detect anemia is
   a. type and cross match.
   b. hemoglobin.
   c. white blood cell count.
   d. prothrombin time.

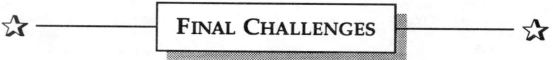

FINAL CHALLENGES

Use a separate sheet of paper to complete this section.

1. Patients with advanced kidney diseases that impair kidney function can become anemic. On the other hand, patients with kidney tumors can develop polycythemia. Explain. (Hint: Tumors can cause tissue overactivity)

2. When track athletes from low altitudes are going to compete at a high altitude, they try to spend a few weeks training at a high altitude. Explain how high altitude training will improve the athlete's ability to compete.

3. During pregnancy the developing fetus manufactures many new red blood cells. What dietary precautions should the mother take to prevent anemia in herself and the fetus?

4. Bob Aker is treating his arthritis by taking aspirin. Prior to having a tooth removed, his dentist asks him to stop taking the aspirin. Explain the dentist's request.

5. Would a transfusion reaction result if the donor is type AB and the recipient is type O?

ANSWERS TO CHAPTER 11

CONTENT LEARNING ACTIVITY

Introduction
1. Oxygen; 2. Carbon dioxide; 3. Body temperature;
4. pH; 5. Microorganisms; 6. Clotting

Plasma
1. Water; 2. Albumin; 3. Globulins; 4. Serum

Formed Elements
1. Red blood cells; 2. White blood cells; 3. Platelets

Red blood cells
A. 1. Hemoglobin; 2. Heme; 3. Folic acid and
   vitamin B_{12}; 4. Erythropoietin
B. 1. Macrophages; 2. Iron; 3. Bilirubin; 4. Jaundice;
   5. Bilirubin
C. 1. Hemoglobin; 2. Heme; 3. Globin; 4. Iron;
   5. Bilirubin

White Blood Cells and Platelets
1. Neutrophil; 2. Monocyte; 3. Lymphocyte; Basophil
and eosinophil; Platelets

Preventing Blood Loss
A. 1. Platelet plug; 2. Calcium; 3. Clotting factors;
   4. Prothrombin activator; 5. Thrombin; 6. Fibrin;
   7. Clot; 8. Liver; 9. Vitamin K
B. 1. Anticoagulant; 2. Aspirin; 3. Thrombus;
   4. Embolus; 5. Plasmin; 6. Streptokinase and
   tissue plasminogen activator

Blood Grouping
A. 1. Transfusion; 2. Antigens; 3. Antibodies;
   4. Agglutination; 5. Hemolysis; 6. Blood groups
B. 1. Type A blood; 2. Type O blood; 3. Recipient;
   4. Type O blood; 5. No reaction; 6. Transfusion
   reaction

Understanding Rh Blood Groups
1. Rh-positive; 2. Hemolytic disease of the newborn;
3. Rh antibodies; 4. Rh-positive; 5. Rh-negative;
6. Anti-Rh_O(D); 7. Bilirubin; 8. Rh-negative

Diagnostic Blood Tests
1. Type and cross match; 2. Complete blood count;
3. Red blood cell count; 4. Hematocrit; 5. White
blood cell count; 6. White blood cell differential
count; 7. Platelet count; 8. Prothrombin time;
9. Blood chemistry

Some Disorders of the Blood
A. 1. Aplastic anemia; 2. Iron deficiency anemia;
   3. Pernicious anemia; 4. Hemorrhagic anemia;
   5. Hemolytic anemia; 6. Thalassemia; 7. Sickle
   cell anemia
B. 1. Hemophilia; 2. Leukemia; 3. Septicemia;
   4. Malaria; 5. Mononucleosis; 6. AIDS and
   hepatitis

QUICK RECALL

1. Functions of blood: 1. Transport of substances to
cells (e.g., oxygen, hormones); 2. Transport of
substances from cells (e.g., carbon dioxide); 3.
Maintenance of body temperature; 4. Maintenance
of fluid, electrolyte, and pH balance; 5. Protection
against microorganisms; 6. Clots to prevent fluid and
cell loss

2. Oxygen binds to iron, which is contained with heme
molecules. Heme and globins form hemoglobin.

3. The iron in heme is reused to produce new
hemoglobin, the rest of heme becomes bilirubin
which is excreted in bile; globin is broken down into
amino acids that are used to produce proteins.

4. The kidneys release erythropoietin, which stimulates
red bone marrow to increase red blood cell
production.

5. Neutrophil - phagocytosis; basophil - inflammation;
eosinophil - inflammation; lymphocyte - immune
response, e.g., antibody production; monocyte -
becomes macrophage, a large phagocytic cell

6. Platelet plug; release chemicals necessary for the
chemical reactions of clotting

7. Prothrombin activator converts prothrombin to
thrombin. Thrombin converts fibrinogen to fibrin.

8. Antigens on the surface of red blood cells react with
antibodies in the plasma. As a result red blood cells
clump (agglutinate) or rupture (hemolysis).

165

1. E. Blood performs all the functions listed.

2. D. Platelets are formed elements.

3. D. Red blood cells do not have a nucleus. It is lost during the formation of the red blood cells. Red blood cells are the most numerous formed element, they are disk-shaped cells, and they are produced in red marrow.

4. D. Hemoglobin consists of the red-pigment heme and protein globins. Iron in the heme binds to and transports oxygen. Hemoglobin transports 97% of the oxygen in blood, and plasma transport 3%.

5. B. Globin is broken down into amino acids that can be used to synthesize proteins. Heme is broken down into bilirubin and iron is reused to form new hemoglobin molecules.

6. C. Decreased blood oxygen levels stimulates the release of erythropoietin from the kidneys. The erythropoietin stimulates red blood cell production.

7. C. The most common white blood cells are neutrophils. They phagocytize microorganisms and other substances.

8. B. Monocytes give rise to macrophages. They are the largest-sized white blood cells. Neutrophils typically accumulated to form pus at sites of infection.

9. D. Exposure to collagen fibers in a torn blood vessel results in chemical reactions that lead to the production of prothrombin activator. Prothrombin activator converts prothrombin to thrombin. Thrombin converts fibrinogen into fibrin, the clot.

10. D. Plasminogen is converted into plasmin, which breaks down the clot. Anticoagulants prevent clot formation. Clotting factors, when activated, result in the formation of a clot. Fibrinogen becomes fibrin, which forms the clot.

11. C. A transfusion reaction occurs when type B blood is given to a person with type A blood. Type B blood has B antigens and A antibodies, whereas type A blood has A antigens and B antibodies. A transfusion reaction occurs because the type B antibodies in the recipient react with the B antigens in the donor's blood.

12. C. The anti-Rh_o(D) is antibodies that bind to any Rh-positive red blood cells from the fetus that may have entered the mother's blood. This prevents the mother from becoming sensitized to the Rh-positive antigen.

13. B. The hemoglobin test determines the amount of hemoglobin in a given volume of blood. Anemia can be caused by reduced amounts of hemoglobin.

 # FINAL CHALLENGES

1. Impairment of kidney function results in decreased erythropoietin production. Thus red blood cell production decreases and anemia results. If a tumor causes overproduction of erythropoietin, polycythemia results.

2. When a person accustomed to living at sea level moves to a higher altitude, there is a decreased ability to get enough oxygen. This can seriously impair athletic performance. The low blood oxygen, however, stimulates erythropoietin release from the kidneys. The erythropoietin stimulates increased red blood cell production in red bone marrow. After enough red blood cells are produced, they increase the ability of the blood to pick up oxygen, and athletic performance improves.

3. The mother should include adequate amounts of vitamin B_{12} and folic acid (to ensure red blood cell production), iron (to ensure hemoglobin production), and vitamin K (to ensure proper blood clotting).

4. Aspirin inhibits prostaglandin synthesis by platelets. Consequently, the activation of platelets is inhibited. Platelet activation is necessary for platelet plug formation and for the release of chemicals such as calcium, which are required for clot formation. A person taking aspirin is more likely to bleed, and this would not be recommended for someone about to have a tooth extracted.

5. Type AB blood has A and B antigens, but no A or B antibodies, whereas type O blood has no A or B antigens, but does have A and B antibodies. If type AB blood is given to a person with type O blood, the A and B antibodies in the type O blood react with the A and B antigens in the type AB blood to cause a transfusion reaction.

The Heart

FOCUS: The heart is a muscular pump composed of two atria, which pump blood to the ventricles, and two ventricles, which pump blood to the body and lungs. The heart is surrounded by pericardium, which reduces friction and anchors the heart in the thoracic cavity. Tricuspid, bicuspid, and semilunar valves ensure one-way flow of blood through the heart, and the heart sounds are produced as these valves close. Cardiac muscle cells are elongated, branching cells that appear striated. Cardiac muscle cells function as a unit, so that the heart can operate as a pump. Specialized cardiac muscle cells originate action potentials which produce the rhythmic contraction (systole) and relaxation (diastole) of the heart. One diastole and the next systole represent one cardiac cycle. Both intrinsic (i.e., Starling's law of the heart) and extrinsic (i.e., baroreceptor reflex, chemoreceptors, emotions, and hormones) mechanisms control heart function.

CONTENT LEARNING ACTIVITY

Size, Form, and Location of the Heart

"*It is important to know the location of the heart, for use of a stethoscope, recording an ECG, or CPR.***"**

Using the terms provided, complete the following statements:

Apex   Lungs
Base   Right
Cone   Thoracic
Left

The adult heart has the shape of a blunt _(1)_, and is about the size of a closed fist. It is located in the _(2)_ cavity between the _(3)_. The blunt, rounded point of the cone is the _(4)_, and the larger, flat portion at the opposite end of the cone is the _(5)_. The apex is the most inferior part of the heart and it is directed to the _(6)_.

1. _____

2. _____

3. _____

4. _____

5. _____

6. _____

Pericardial Sac

"*The pericardial sac surrounds the heart and anchors it within the thoracic cavity.***"**

Match these terms with the
correct statement or definition:

Parietal pericardium   Pericarditis
Pericardial cavity   Serous pericardium
Pericardial fluid   Visceral pericardium

_____   1. Delicate layer of squamous epithelial cells that lines the pericardial sac.

_____   2. Serous pericardium that covers the heart surface; epicardium.

_____   3. Space between the visceral and parietal pericardia.

_____   4. Fluid in the pericardial cavity that helps reduce friction as the heart moves.

_____   5. Inflammation of the pericardium.

Heart Chambers and Valves

"*The heart is a muscular pump consisting of four chambers, two atria and two ventricles.***"**

A. Match these terms with the
correct statement or definition:

Aorta   Pulmonary trunk
Interatrial septum   Pulmonary veins
Interventricular septum   Venae cavae and coronary sinus

_____   1. Wall that separates the right and left atria.

_____   2. Blood vessels entering the right atrium.

_____   3. Blood vessels entering the left atrium.

_____   4. Blood vessel leaving the right ventricle.

_____   5. Blood vessel leaving the left ventricle.

B. Match these terms with the
correct statement or definition:

Bicuspid (mitral) valve   Semilunar valves
Chordae tendineae   Tricuspid valve
Papillary muscles

_____   1. Valve between the right atrium and right ventricle.

_____   2. Valve between the left atrium and left ventricle.

_____   3. Cone-shaped muscular pillars in each ventricle.

_____   4. Connective tissue strings connecting papillary muscles with the cusps of the tricuspid and bicuspid valves.

_____   5. Valves with three pocketlike cusps, located in the aorta and pulmonary trunk.

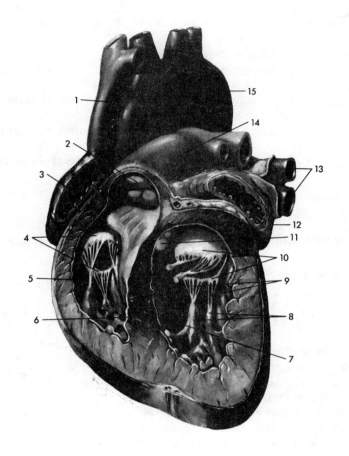

C. Match these terms with the correct parts labeled in Figure 12-1:

Aorta
Aortic semilunar valve
Bicuspid (mitral) valve
Chordae tendineae
Interventricular septum
Left atrium
Left ventricle

Papillary muscles
Pulmonary semilunar valve
Pulmonary trunk
Pulmonary veins
Right atrium
Right ventricle
Superior vena cava
Tricuspid valve

1. _____

2. _____

3. _____

4. _____

5. _____

6. _____

7. _____

8. _____

9. _____

10. _____

11. _____

12. _____

13. _____

14. _____

15. _____

Heart Wall

"_The heart wall is composed of three layers of tissue._**"**

Match these terms with the
correct statement or definition:

Endocardium   Myocardium
Epicardium

_____ 1. Visceral pericardium; forms the smooth outer surface of the heart.

_____ 2. The middle layer of the heart; consists of cardiac muscle cells.

_____ 3. The smooth inner surface of the heart chambers.

Cardiac Muscle

"_Cardiac muscle cells are elongated, branching, striated cells with one or occasionally two nuclei._**"**

Using the terms provided, complete the following statements:

ATP   Mitochondria
Intercalated disks   Oxygen

The cardiac muscle cells are bound to each other by specialized
cell-to-cell contacts called _(1)_, where action potentials pass
from cell to cell and cause all of the cardiac muscle cells to
contract. The energy for cardiac muscle contraction is provided
by _(2)_. Cardiac muscle cells have many _(3)_, where ATP is
produced at a rapid enough rate to sustain muscle contraction.
An extensive capillary network provides an adequate
_(4)_ supply to the cells, because, unlike skeletal muscle, cardiac
muscle cannot develop a significant oxygen debt.

1. _____

2. _____

3. _____

4. _____

Blood Flow Through the Heart

"_The pumping action of the heart depends on the alternating contraction of the two atria and_**"**
the two ventricles.

A. Match these terms with the
correct statement or definition:

Cardiac cycle   Systole
Diastole

_____ 1. Relaxation ventricles.

_____ 2. Contraction of the ventricles.

_____ 3. Diastole and the next systole.

B. Using the terms provided, complete the following statements:

Aortic semilunar   Pulmonary trunk
Bicuspid (mitral)   Pulmonary veins
Left atrium   Right atrium
Pulmonary arteries   Right ventricle
Pulmonary semilunar   Tricuspid

Blood enters the _(1)_ from all the parts of the body except the lungs. Blood then flows into the _(2)_, which completes filling as the right atrium contracts. Contraction of the right ventricle pushes blood against the _(3)_ valve, forcing it closed, and against the _(4)_ valve, forcing it open, and allowing blood to flow into the _(5)_. The _(6)_ carry blood to the lungs, where carbon dioxide is released, and oxygen picked up. Blood returning from the lungs enters the _(7)_ through the four _(8)_. Blood in the left atrium pushes the _(9)_ valve open, and contraction of the left atrium completes filling of the left ventricle. Contraction of the left ventricle pushes blood against the _(10)_ valve, closing it, and against the _(11)_ valve, opening it and allowing blood to enter the aorta.

1. _____
2. _____
3. _____
4. _____
5. _____
6. _____
7. _____
8. _____
9. _____
10. _____
11. _____

Understanding Coronary Circulation

"Cardiac muscle in the wall of the heart is thick and metabolically very active.**"**

Match these terms with the correct statement or definition:

Angioplasty   Coronary bypass
Angina pectoris   Coronary thrombosis
Atherosclerotic lesions Infarct
Cardiac veins   Thrombus
Coronary arteries

_____ 1. Vessels that supply blood to the wall of the heart.

_____ 2. Vessels that drain blood from the cardiac muscle and empty into the coronary sinus.

_____ 3. A blood clot.

_____ 4. A heart attack.

_____ 5. Region of dead heart tissue.

_____ 6. Thickenings in the walls of arteries containing deposits high in cholesterol and other lipids.

_____ 7. Pain in the chest and left arm that occurs when the heart is deprived of an adequate blood supply.

_____ 8. Treatment in which a small balloon is used to open blockage in a coronary artery.

_____ 9. Surgery that uses healthy blood vessels from another part of the body to bypass obstructions in the coronary arteries.

Conduction System of the Heart

"The heart is mainly regulated by specialized cardiac muscle cells in the wall of the heart."

A. Match these terms with the correct statement or definition:

Atrioventricular bundle
AV node
Bundle branches

Purkinje fibers
SA node

_____ 1. Located in the upper wall of the right atrium; initiates contraction of the heart; the pacemaker.

_____ 2. Located in the lower portion of the right atrium; slows the rate of action potential conduction.

_____ 3. Conducting cells that arise from the AV node; rapid action potential conduction occurs here.

_____ 4. Right and left subdivisions of the atrioventricular bundle.

_____ 5. Numerous small bundles of conducting tissue that extend to the cardiac muscle of the ventricles.

B. Match these terms with the correct parts labeled in Figure 12-2:

Atrioventricular bundle
AV node
Bundle branches
Purkinje fibers
SA node

1. _____

2. _____

3. _____

4. _____

5. _____

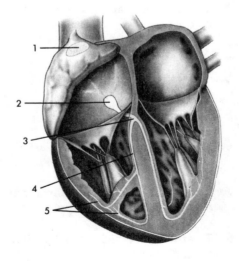

C. Match these terms with the
correct statement or definition:

Defibrillation   Fibrillation
Ectopic beat

_____ 1. Action potentials originating in an area of the heart other than the
   SA node.

_____ 2. Condition in which many small portions of the heart contract
   rapidly and independently of each other.

_____ 3. Application of a strong electrical shock to the chest region to
   depolarize all the cardiac muscle cells at the same time.

☞ Because action potentials originate from the SA node automatically, without stimulation from
the nervous or endocrine systems, the heart is said to be autorhythmic.

Electrocardiogram

❝The record of electrical changes resulting from action potentials in cardiac muscle**❞**
is an electrocardiogram.

A. Match these terms with the
correct statement or definition:

P-Q (P-R) interval   Q-T interval
P wave   T wave
QRS complex

_____ 1. Record of action potentials during depolarization of the atria.

_____ 2. Record of action potentials from depolarization of the ventricles.

_____ 3. Record of repolarization of the ventricles.

_____ 4. Time during which the atria contract and begin to relax.

_____ 5. Length of time required for ventricular depolarization and
   repolarization.

B. Match these terms with
   the correct parts labeled
   in Figure 12-3:

   P-Q (P-R) interval
   P wave
   QRS complex
   Q-T interval
   T wave

1. _____

2. _____

3. _____

4. _____

5. _____

Heart Sounds

❝*There are two main heart sounds.***❞**

Match these terms with the
correct statement or definition:

First heart sound   Second heart sound
Incompetent valve   Stenosed valve
Murmur

_____ 1. Results from the closure of tricuspid and bicuspid valves.

_____ 2. Results from the closure of semilunar valves.

_____ 3. Caused by a faulty valve; swishing sound occurs before or after valve
                            closure.

_____ 4. Valve that allows blood to leak through when it is closed; swishing
                            sound occurs immediately after valve closure.

_____ 5. Narrowed valve, which causes swishing sound before valve closure

Regulation of Heart Function

❝*There are a number of mechanisms that modify heart rate and stroke volume.***❞**

Match these terms with the
correct statement or definition:

Cardiac output   Stroke volume
Heart rate

_____ 1. Volume of blood pumped by the ventricles of the heart each minute
                            (stroke volume x heart rate).

_____ 2. Volume of blood pumped per ventricle each time the heart contracts.

_____ 3. Number of times the heart contracts each minute.

Intrinsic Regulation of the Heart

66_Intrinsic regulation of the heart refers to mechanisms contained within the heart itself._**99**

Using the terms provided, complete the following statements:

Decreased   Increased
Starling's law of the heart

The heart fills to a greater volume when the amount of blood that returns to the heart is _(1)_. The greater volume stretches the cardiac muscle fibers, which contract with _(2)_ force. Consequently, a(n) _(3)_ volume of blood is pumped from the heart, resulting in _(4)_ stroke volume. The stretch also causes a slightly _(5)_ heart rate. Therefore, if the volume of blood entering the heart is increased, cardiac output is _(6)_, whereas, if the volume of blood entering the heart is decreased, cardiac output is _(7)_. This direct relationship between the volume of blood entering the heart and cardiac output is called _(8)_.

1. _____

2. _____

3. _____

4. _____

5. _____

6. _____

7. _____

8. _____

Extrinsic Regulation of the Heart

66_Extrinsic regulation of the heart refers to both nervous and hormonal regulation of the heart._**99**

A. Match these terms with the correct statement or definition:

Baroreceptors   Cardioregulatory center
Chemoreceptors

_____

_____

_____

1. Sensory receptors sensitive to the stretch of the walls of the aorta and internal carotid arteries.

2. Part of the medulla oblongata that receives and integrates action potentials from baroreceptors.

3. Sensory receptors in the medulla oblongata sensitive to changes in pH and carbon dioxide levels.

B. Complete each statement by providing the missing word:

Decrease(s)
Increase(s)

_____

_____

_____

_____

_____

1. Increased blood pressure causes stretching of baroreceptors, which increases parasympathetic stimulation and _____ heart rate.

2. Excitement, anxiety, or anger increases sympathetic stimulation of the heart, which _____ cardiac output.

3. Epinephrine and norepinephrine from the adrenal medulla _____ heart rate and stroke volume.

4. Decrease in pH and an increase in carbon dioxide _____ sympathetic stimulation of the heart.

5. Decreased body temperature _____ heart rate.

The autonomic nervous system innervates the SA node of the heart; sympathetic stimulation of the heart increases heart rate and stroke volume, whereas parasympathetic stimulation decreases heart rate.

Conditions and Diseases Affecting the Heart

"Many conditions and diseases can affect the ability of the heart to supply blood to the body."

A. Match these terms with the correct statement or definition:

Endocarditis   Pericarditis
Myocarditis   Rheumatic heart disease

_____ 1. Inflammation of the lining of the heart cavities.

_____ 2. Inflammation of the serous membrane of the pericardium.

_____ 3. Endocardial inflammation that can lead to heart valve damage; results from immune reaction caused by streptococcal infection.

B. Match the condition with the correct statement or definition:

Aortic or pulmonary stenosis   Patent ductus arteriosus
Cyanosis   Septal defect
Heart failure

_____ 1. Hole in one of the septums between the right and left sides of the heart.

_____ 2. Failure of a blood vessel between the aorta and pulmonary trunk to close after birth.

_____ 3. Narrowed vessel or valve; can cause peripheral or pulmonary edema.

_____ 4. Blueness of the skin caused by low oxygen levels in the blood.

_____ 5. Progressive weakening of the heart muscle, resulting in edema.

C. Using the terms provided, complete the following statements:

Decreased
Increased

Proper nutrition is important in reducing the risk of heart disease. In a recommended diet, saturated fat and cholesterol should be _(1)_ , and fiber, whole grains, fruits and vegetables should be _(2)_ . Total food intake should be limited, and salt intake _(3)_ . Tobacco use and excessive use of alcohol also lead to _(4)_ risk of heart disease. Chronic stress and lack of physical exercise lead to a(n) _(5)_ risk of cardiovascular disease; remedies include relaxation and aerobic exercise. Hypertension is _(6)_ systemic blood pressure, and must be controlled by diet, exercise, or medication.

1. _____

2. _____

3. _____

4. _____

5. _____

6. _____

D. Match the term with the correct statement or definition:

Anticoagulants
Beta blockers
Calcium channel blockers
Digitalis
Heart lung machine
Heart transplant
Nitroglycerin
Pacemaker

_____ 1. Medication that slows and strengthens heart contractions.

_____ 2. Medication that dilates coronary blood vessels and increases blood flow to cardiac muscle; used to treat angina pectoris.

_____ 3. Medication that reduces the rate and strength of contraction; blocks beta-adrenergic receptors; often used to treat rapid heart rate or arrhythmias.

_____ 4. Medication that reduces the rate at which calcium ions diffuse into cardiac muscle cells; reduces hypertension, arrhythmia, and tachycardia.

_____ 5. Prevent clot formation; include aspirin.

_____ 6. Instrument that provides an electrical stimulus to the heart.

_____ 7. Machine that oxygenates and removes carbon dioxide from the blood; used during surgery on the heart and lungs.

_____ 8. For this procedure, immune characteristics of donor and recipient must be matched.

☞ Artificial hearts have been used experimentally, but currently do not function well enough to allow a high quality of life.

QUICK RECALL

1. Name the four valves that regulate the direction of blood flow in the heart, and give their location.

2. Give the two nodes of the conducting system of the heart and their functions.

3. State the cause of the P wave, the QRS complex, and the T wave of the ECG. Name the contraction event associated with each wave.

4. List the two normal heart sounds, and give the reason for each.

5. Define Starling's law of the heart.

6. List the effects of increased blood pressure, epinephrine, and greatly increased blood carbon dioxide levels (emergency conditions) on the heart.

MASTERY LEARNING ACTIVITY

Place the letter corresponding to the correct answer in the space provided.

_____ 1. Which of the following structures carry blood to the right atrium?
   a. coronary sinus
   b. superior vena cava
   c. inferior vena cava
   d. all of the above

_____ 2. The pericardial cavity
   a. is located between the parietal and visceral pericardia.
   b. is lined with connective tissue.
   c. is filled with air.
   d. all of the above

_____ 3. The valve located between the right atrium and the right ventricle is the
   a. aortic semilunar valve.
   b. pulmonary semilunar valve.
   c. tricuspid valve.
   d. bicuspid (mitral) valve.

_____ 4. The papillary muscles
   a. are attached to the chordae tendineae.
   b. are found in the atria.
   c. are located on the external surface of the heart.
   d. are attached to the semilunar valves.

_____ 5. The thickest part of the heart wall is
   a. epicardium
   b. myocardium
   c. pericardium
   d. endocardium

_____ 6. Given the following blood vessels:
   1. aorta
   2. inferior vena cava
   3. pulmonary trunk
   4. pulmonary vein

Choose the arrangement that lists the vessels in the order a red blood cell would encounter them as it returned to the heart.
   a. 1,3,4,2
   b. 2,3,4,1
   c. 2,4,3,1
   d. 3,2,1,4

_____ 7. Which of the following correctly describes conditions during the cardiac cycle?
   a. When the right ventricle contracts, the bicuspid valve is pushed open.
   b. The ventricles are mostly filled before the atria contract.
   c. Contraction of the atria closes the aortic and pulmonary semilunar valves.
   d. During diastole, the bicuspid and tricuspid valves close.
   e. all of the above

_____ 8. Given the following structures of the conduction system of the heart:
   1. atrioventricular bundle
   2. AV node
   3. bundle branches
   4. Purkinje fibers
   5. SA node

   Choose the arrangement that lists the structures in the order an action potential would pass through them.
   a. 2,5,1,3,4
   b. 2,5,3,1,4
   c. 2,5,4,1,3
   d. 5,2,1,3,4
   e. 5,2,4,3,1

_____ 9. A T wave represents
   a. depolarization of the ventricles.
   b. repolarization of the ventricles.
   c. depolarization of the atria.
   d. repolarization of the atria.

_____ 10. The "dupp" sound (second heart sound) is caused by
   a. the closing of the bicuspid and tricuspid valves.
   b. the closing of the semilunar valves.
   c. blood rushing out of the ventricles.
   d. the filling of the ventricles.
   e. ventricular contraction.

_____ 11. A murmur is the result of
   a. the heart skipping a beat.
   b. a faulty valve.
   c. digestive disturbances.
   d. a faulty SA node.

_____ 12. Cardiac output is defined as
   a. blood pressure times heart rate.
   b. heart rate times 60.
   c. heart rate times stroke volume.
   d. stroke volume times blood pressure

_____ 13. Increased blood entering the heart results in increased
   a. stroke volume.
   b. heart rate.
   c. cardiac output.
   d. all of the above

_____ 14. Because of the baroreceptor reflex, when blood pressure decreases below normal, you would expect
   a. heart rate to decrease.
   b. stroke volume to decrease.
   c. increased parasympathetic stimulation of the heart.
   d. blood pressure to return to normal.
   e. all of the above

_____ 15. A decrease in blood pH and an increase in blood carbon dioxide levels results in
   a. increased heart rate.
   b. increased stroke volume.
   c. increased sympathetic stimulation of the heart.
   d. all of the above

Use a separate sheet of paper to complete this section.

1. The Jarvik-7 artificial heart is designed as a replacement for the ventricles. Explain why it is more important to replace the ventricles than the atria of the heart.

2. During an experiment in a physiology laboratory a student named C. Saw was placed on a table that could be tilted. The instructor asked the students to predict what would happen to C. Saw's heart rate if the table were tilted so that her head were lower than her feet. Some students predicted an increase in heart rate, and others claimed it would decrease. Can you explain why both predictions might be true?

3. After C. Saw was tilted so that her head was lower than her feet for a few minutes, the table was tilted so that her head was higher than her feet. Predict the effect this change would have on C. Saw's heart rate.

4. In most tissues peak blood flow occurs during systole and decreases during diastole. However, in heart tissue the opposite is true and peak blood flow occurs during diastole. Explain why this difference occurs.

5. Mr. Calch Potatoe had his heart rate and cardiac output measured. After two months of aerobic training, the measurements were repeated. It was found that his heart rate had decreased, but his cardiac output remained the same for many activities. Explain these results.

6. In the conducting system of the heart, action potentials originate in the SA node (pacemaker), travel through the atria to the AV node, and then through the AV bundle, bundle branches, and the Purkinje fibers, which pass around the apex of the heart. Because of this arrangement, ventricular contraction begins at the apex of the heart. Explain why it is more efficient for contraction of the ventricles to begin at the apex than at the base.

ANSWERS TO CHAPTER 12

CONTENT LEARNING ACTIVITY

Size, Form, and Location of the Heart.
　1. Cone; 2. Thoracic; 3. Lungs; 4. Apex; 5. Base; 6. Left
Pericardial Sac
　1. Serous pericardium; 2. Visceral pericardium;
　3. Pericardial cavity; 4. Pericardial fluid;
　5. Pericarditis
Heart Chambers and Valves
　A. 1. Interatrial septum; 2. Venae cavae and coronary sinus; 3. Pulmonary veins; 4. Pulmonary trunk; 5. Aorta
　B. 1. Tricuspid valve; 2. Bicuspid (mitral) valve; 3. Papillary muscles; 4. Chordae tendineae; 5. Semilunar valves

　C. 1. Superior vena cava; 2. Pulmonary semilunar valve; 3. Right atrium; 4. Tricuspid valve; 5. Right ventricle; 6. Interventricular septum; 7. Left ventricle; 8. Papillary muscles; 9. Chordae tendineae; 10. Bicuspid (mitral) valve; 11. Aortic semilunar valve; 12. Left atrium; 13. Pulmonary veins; 14. Pulmonary trunk; 15. Aorta
Heart Wall
　1. Epicardium; 2. Myocardium; 3. Endocardium
Cardiac Muscle
　1. Intercalated disks; 2. ATP; 3. Mitochondria; 4. Oxygen
Blood Flow Through the Heart
　A. 1. Diastole; 2. Systole; 3. Cardiac cycle

B. 1. Right atrium; 2. Right ventricle; 3. Tricuspid;
   4. Pulmonary semilunar; 5. Pulmonary trunk;
   6. Pulmonary arteries; 7. Left atrium;
   8. Pulmonary veins; 9. Bicuspid (mitral);
   10. Bicuspid (mitral); 11. Aortic semilunar valve

Coronary Circulation
   1. Coronary arteries; 2. Cardiac veins; 3. Thrombus;
   4. Coronary thrombosis; 5. Infarct; 6. Atherosclerotic
   lesions; 7. Angina pectoris; 8. Angioplasty;
   9. Coronary bypass

Conduction System of the Heart
   A. 1. SA node; 2. AV node; 3. Atrioventricular
      bundle; 4. Bundle branches; 5. Purkinje fibers
   B. 1. SA node; 2. AV node; 3. Atrioventricular
      bundle; 4. Bundle branches; 5. Purkinje fibers
   C. 1. Ectopic beat; 2. Fibrillation; 3. Defibrillation

Electrocardiogram
   A. 1. P wave; 2. QRS complex; 3. T wave; 4. P-Q (P-R)
      interval; 5. Q-T interval
   B. 1. P wave; 2. QRS complex; 3. T wave; 4. P-Q (P-R)
      interval; 5. Q-T interval

Heart Sounds
   1. First heart sound; 2. Second heart sound;
   3. Murmur; 4. Incompetent valve; 5. Stenosed valve

Regulation of Heart Function
   1. Cardiac output; 2. Stroke volume; 3. Heart rate
Intrinsic Regulation of the Heart
   1. Increased; 2. Increased; 3. Increased;
   4. Increased; 5. Increased; 6. Increased;
   7. Decreased; 8. Starling's law of the heart
Extrinsic Regulation of the Heart
   A. 1. Baroreceptors; 2. Cardioregulatory center;
      3. Chemoreceptors
   B. 1. Decreases; 2. Increases; 3. Increase;
      4. Increases; 5. Decreases
Conditions and Diseases Affecting the Heart
   A. 1. Endocarditis; 2. Pericarditis; 3. Rheumatic
      heart disease
   B. 1. Septal defect; 2. Patent ductus arteriosus;
      3. Aortic or pulmonary stenosis; 4. Cyanosis;
      5. Heart failure
   C. 1. Decreased; 2. Increased; 3. Decreased;
      4. Increased; 5. Increased; 6. Increased
   D. 1. Digitalis; 2. Nitroglycerin; 3. Beta blockers;
      4. Calcium channel blockers; 5. Anticoagulants;
      6. Pacemaker; 7. Heart lung machine; 8. Heart
      transplant

QUICK RECALL

1. Tricuspid valve: between right atrium and right
   ventricle; bicuspid (mitral) valve: between left
   atrium and left ventricle; pulmonary semilunar
   valve: in the pulmonary trunk; aortic semilunar
   valve: in the aorta
2. SA node: pacemaker of the heart; AV node: slows
   action potentials and allows atria to move blood into
   the ventricles
3. P wave: caused by depolarization of the atria, atrial
   systole; QRS complex: caused by depolarization of
   the ventricles, ventricular systole; T wave: caused by
   repolarization of the ventricles, ventricular diastole.

4. First heart sound: closing of tricuspid and bicuspid
   valves; second heart sound: closing of semilunar
   valves.
5. Starling's law of the heart: If the volume of blood
   entering the heart increases, cardiac output
   increases; if the volume of blood entering the heart
   decreases, cardiac output decreases.
6. Increased blood pressure: decreased heart rate and
   stroke volume from baroreceptor reflex;
   epinephrine: increased heart rate and stroke
   volume; greatly increased blood carbon dioxide
   levels (emergency conditions): increased heart rate
   and stroke volume from sympathetic stimulation

MASTERY LEARNING ACTIVITY

1. D. The coronary sinus, inferior vena cava, and
   superior vena cava all carry blood to the right atrium.

2. A. The pericardial cavity is located between the
   visceral and parietal pericardia. It is lined with
   serous pericardium which secretes pericardial fluid.
   Pericardial fluid reduces friction as the heart moves
   within the pericardial sac.

3. C. The tricuspid valve is located between the right
   atrium and the right ventricle. The bicuspid valve is
   located between the left atrium and left ventricle.
   The semilunar valves are in the aorta and the
   pulmonary trunk.

4. A. The papillary muscles are found in the
   ventricles and are attached to the chordae
   tendineae, which in turn are attached to the cusps of
   the tricuspid and bicuspid valves.

5. B.   The myocardium is the middle layer of the heart wall, is composed of cardiac muscle, and comprises the bulk of the heart.

6. B.   The red blood cell would pass through the following structures: inferior vena cava, right atrium, right ventricle, pulmonary trunk, pulmonary arteries, lungs, pulmonary veins, left atrium, left ventricle, aorta.

7. B.   Both ventricles are mostly filled before atrial contraction occurs by blood flowing from the atria as the ventricles relax. Contraction of the atria completes ventricular filling. During systole, the tricuspid and bicuspid valves are pushed closed. Atrial contraction does not affect the aortic and pulmonary semilunar valves. During diastole, both bicuspid and tricuspid valves are open.

8. D.   Action potentials originate in the SA node, pass to the AV node, the atrioventricular bundle, the bundle branches, and the Purkinje fibers to the myocardium of the ventricles.

9 B.   T waves represent repolarization of the ventricles. The QRS complex represents depolarization of the ventricles, and P waves represent depolarization of the atria.

10. B.   The "dupp" (second heart sound) is caused by the closing of the semilunar valves.

11. B.   A murmur is an abnormal heart sound accompanying an incompetent valve (sound of blood leaking through) or stenosed valve (sound of blood passing through a narrowed valve opening).

12. C.   Cardiac output is heart rate .times stroke volume.

13. D.   Increased blood entering the heart results in increased stroke volume and heart rate. The increased stroke volume and heart rate result in increased cardiac output.

14. D.   When blood pressure decreases, the baroreceptor reflex results in sympathetic stimulation, which causes an increase in heart rate and stroke volume. Consequently, blood pressure increases (returns to normal).

15. D.   A decrease in blood pH and an increase in blood carbon dioxide stimulate chemoreceptors in the medulla oblongata of the brain. This results in sympathetic stimulation of the heart, increasing heart rate and stroke volume.

★ FINAL CHALLENGES ★

1. In artificial heart implants it is most important to replace the ventricles, which pump blood to the lungs and to the body. Because most ventricular filling takes place before atrial contraction, the heart can function fairly well without the pumping action of the atria.

2. A decrease in heart rate could be expected due to the baroreceptor reflex, because tilting could increase blood pressure in the carotid arteries. On the other hand, an increase in heart rate could occur because of stretch of the SA node, because tilting would increase the blood returning to the right atrium.

3. After being tilted, the upright position would cause a decrease in pressure in the carotid arteries. Through the baroreceptor reflex, this would cause an increase in heart rate.

4. The heart is in a contracted state when systolic pressure is generated. The contracted state creates external pressure on the coronary blood vessels, and reduces the blood flow. When diastole occurs, external pressure on the coronary vessels decreases, allowing more blood to flow through them.

5. In trained and untrained individuals, cardiac output is essentially the same because cardiac output is a measure of the amount of blood delivery necessary to perform a given task. In the trained individual, the heart increases in size and the stroke volume increases. Therefore, a lower heart rate combined with the increased stroke volume produces the same cardiac output as the untrained individual with a higher heart rate and lower stroke volume.

6. Contraction of the ventricles, beginning at the apex and moving toward the base of the heart, forces blood out of the inferior ends of the ventricles and toward their outflow vessels—the aorta and pulmonary trunk.

Blood Vessels and Circulation

FOCUS: Blood flows from the heart through the arterial blood vessels to capillaries, and from capillaries back to the heart through veins. The pulmonary circulation transports blood from the heart to the lungs and back to the heart, whereas the systemic circulation carries blood from the heart to the body and back to the heart. Blood pressure is responsible for the movement of blood through the blood vessels, and it is regulated by nervous reflexes and hormones. The kidneys, by controlling blood volume, are the most important long-term regulators of blood pressure.

CONTENT LEARNING ACTIVITY

General Features of Blood Vessel Structure

66 *The makeup of blood vessel walls varies with the diameter of the blood vessel and* 99 *with blood vessel type.*

A. Match these terms with the correct statement or definition:

Tunica externa
Tunica intima
Tunica media

_____ 1. Innermost layer of a blood vessel consisting of simple squamous epithelium (endothelium).

_____ 2. Middle layer of a blood vessel consisting of varying amounts of smooth muscle, elastic fibers, and collagen fibers.

_____ 3. Outer connective tissue layer of a blood vessel.

 Except for the capillaries and the venules, blood vessel walls consist of three relatively distinct layers.

B. Match these terms with the correct statement or definition:

Arterioles   Precapillary sphincters
Artery   Valves
Capillary   Venules
Elastic arteries   Vein
Muscular arteries

_____ 1. General term for the type of vessel that carries blood away from the heart.

_____ 2. Largest arteries; stretch and recoil when blood is pumped into them.

_____ 3. Medium sized and small-diameter arteries; control blood flow to different regions of the body.

_____ 4. Transport blood from small arteries to capillaries; control the amount of blood flowing to specific tissues.

_____ 5. Vessel that consists of only endothelium; site of gas, nutrient, and waste product exchange between blood and tissues.

_____ 6. Regulates blood flow through capillaries.

_____ 7. General term for the type of vessel that carries blood toward the heart.

_____ 8. Carry blood from capillaries to small veins.

_____ 9. Structures in veins that prevent the backflow of blood.

Understanding Blood Vessels

"The walls of all arteries undergo changes as they age."

Match these terms with the correct statement or definition:

Arteriosclerosis   Phlebitis
Atherosclerosis   Varicose veins
Gangrene

_____ 1. Changes in arteries that make them less elastic; hardening of the arteries.

_____ 2. Fatty materials such as cholesterol are deposited in the walls of arteries to form plaques.

_____ 3. Dilated blood vessels in which the valves do not prevent the backflow of blood.

_____ 4. Inflammation of the veins resulting from stagnant blood flow and clotting.

_____ 5. Tissue death and infection of the tissue by bacteria.

Pulmonary Circulation

66 *Pulmonary circulation moves blood from the right ventricle to the lungs and back to the left atrium.* 99

Match these terms with the correct statement or definition:

Pulmonary arteries   Pulmonary veins
Pulmonary trunk

_____ 1. Vessel arising from the right ventricle.

_____ 2. Carry poorly oxygenated blood to the lungs; these two vessels arise from the pulmonary trunk.

_____ 3. Carry oxygenated blood from the lungs to the left atrium; four vessels.

Systemic Circulation: Arteries

66 *Oxygenated blood passes from the left ventricle to the aorta and is distributed to the body.* 99

A. Match these terms with the correct statement or definition:

Abdominal aorta   Descending aorta
Aortic arch   Thoracic aorta
Ascending aorta

_____ 1. Gives rise to the coronary arteries, which supply the heart.

_____ 2. Gives rise to the brachiocephalic, the left common carotid, and the left subclavian arteries.

_____ 3. Longest part of the aorta, running from the aortic arch to the pelvis.

_____ 4. Portion of the aorta between the aortic arch and the diaphragm.

B. Match these arteries with the correct parts labeled in Figure 13-1:

Brachiocephalic artery   1. _____
Left common carotid artery
Left subclavian artery   2. _____
Right common carotid artery
Right subclavian artery   3. _____

                                4. _____

                                5. _____

C. Match these arteries with
   the correct parts labeled
   in Figure 13-2:

   Axillary artery
   Brachial artery
   Radial artery
   Ulnar artery

   1. _____

   2. _____

   3. _____

   4. _____

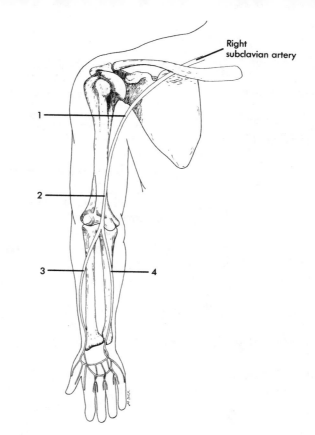

D. Match these arteries with
   the correct parts labeled
   in Figure 13-3:

   Celiac artery
   Common iliac artery
   Inferior mesenteric artery
   Testicular artery
   Renal artery
   Superior mesenteric artery
   Suprarenal artery

   1. _____

   2. _____

   3. _____

   4. _____

   5. _____

   6. _____

   7. _____

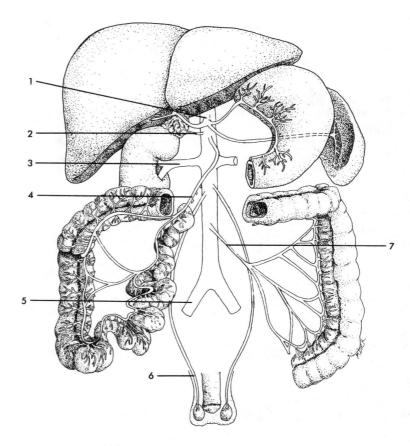

E. **Match these arteries with the correct parts labeled in Figure 13-4:**

External iliac artery
Femoral artery
Internal iliac artery
Popliteal artery

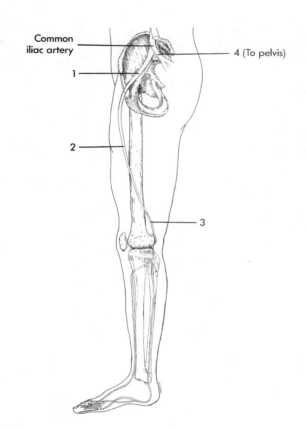

Common iliac artery

4 (To pelvis)

1. _____

2. _____

3. _____

4. _____

Systemic Circulation: Veins

❝*Veins transport deoxygenated blood from the capillaries to the right atrium.***❞**

A. **Match these veins with the correct parts labeled in Figure 13-5:**

Brachiocephalic vein
External jugular vein
Inferior vena cava
Internal jugular vein
Pulmonary veins
Subclavian vein
Superior vena cava

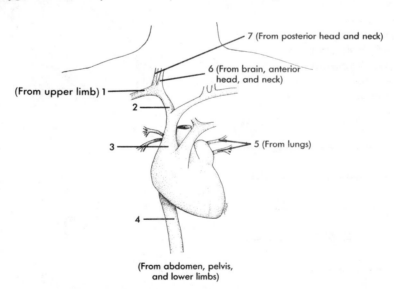

7 (From posterior head and neck)

6 (From brain, anterior head, and neck)

(From upper limb) 1

5 (From lungs)

(From abdomen, pelvis, and lower limbs)

1. _____   4. _____   6. _____

2. _____   5. _____   7. _____

3. _____

B. Match these veins with
   the correct parts labeled
   in Figure 13–6:

   Axillary vein
   Basilic vein
   Brachial vein
   Cephalic vein
   Median cubital vein

Right subclavian vein

5

4 (Deep veins)

1. _____

2. _____

3. _____

4. _____

5. _____

C. Match these veins with
   the correct parts labeled
   in Figure 13–7:

   Common iliac vein
   External iliac vein
   Testicular vein
   Hepatic veins
   Inferior vena cava
   Internal iliac vein
   Renal vein
   Suprarenal vein

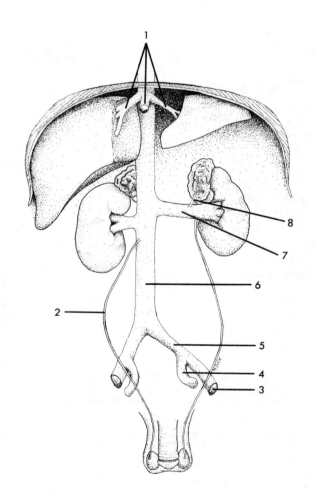

1. _____

2. _____

3. _____

4. _____

5. _____

6. _____

7. _____

8. _____

188

D. Match these veins with
the correct parts labeled
in Figure 13-8:

Gastric vein
Hepatic portal vein
Hepatic veins
Inferior mesenteric vein
Inferior vena cava
Splenic vein
Superior mesenteric vein

1. _____

2. _____

3. _____

4. _____

5. _____

6. _____

7. _____

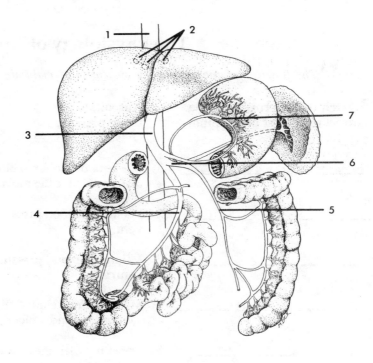

E. Match these veins with
the correct parts labeled
in Figure 13-9:

Femoral vein
Great saphenous vein
Popliteal vein
Small saphenous vein

1. _____

2. _____

3. _____

4. _____

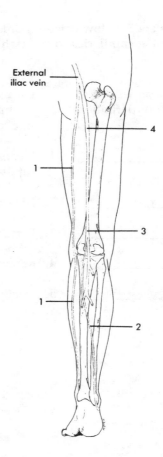

External
iliac vein

The Physiology of Circulation

" *The function of the circulatory system is to maintain adequate blood flow to all tissues.* **"**

A. Match these terms with the
correct statement or definition:

Auscultatory   Pulse
Blood pressure   Systolic pressure
Diastolic pressure

_____ 1. Measure of the force blood exerts against the blood vessel walls;
responsible for the movement of blood through blood vessels.

_____ 2. Maximum blood pressure in the aorta; results from contraction of the
ventricles.

_____ 3. Minimum blood pressure in the aorta; results from relaxation of the
ventricles.

_____ 4. Most common clinical method of determining blood pressure; uses a
stethoscope and a blood pressure cuff.

_____ 5. Pressure when the first turbulent blood flow sounds are heard.

_____ 6. The 80 in a blood pressure measurement of 120/80.

_____ 7. Pressure wave caused by ejection of blood from the left ventricle into
the aorta.

☞ Resistance to flow (friction) causes a drop in blood pressure from the aorta to capillaries, and
from the capillaries to the right atrium.

B. Match these terms with the
correct statement or definition:

Diffusion   Lymphatic capillaries
Edema   Out of capillary
Into capillary

_____ 1. Means by which nutrients and waste products move across the
capillary walls into interstitial spaces.

_____ 2. Direction of fluid movement at the arteriolar end of capillaries.

_____ 3. Direction of fluid movement at the venous end of capillaries.

_____ 4. Removes excess fluid from tissues and returns the fluid to the blood.

_____ 5. Swelling caused by excess fluid accumulation.

Local Control of Blood Vessels

"_Local and nervous control mechanisms match blood flow to the needs of tissues for blood._**"**

Match these terms with the correct statement or definition:

Contraction
Local control
Nervous control

Relaxation
Vasomotor center
Vasomotor tone

_____ 1. Achieved by contraction and relaxation of the precapillary sphincters.

_____ 2. Response of the precapillary sphincters to decreased oxygen or increased carbon dioxide.

_____ 3. Regulates most blood vessels except for capillaries and precapillary sphincters.

_____ 4. Part of the sympathetic nervous system; continually stimulates most blood vessels.

_____ 5. Condition of partial constriction of blood vessels caused by sympathetic stimulation.

_____ 6. Control system that routes blood from the skin and viscera to exercising muscles.

_____ 7. Control system that allows more blood to flow through exercising muscle tissue.

Regulation of Arterial Pressure

"_An adequate blood pressure is required to maintain blood flow through the blood vessels._**"**

Match these terms with the correct statement or definition:

Decreases
Increases

Peripheral resistance

_____ 1. Effect on blood pressure if heart rate or stroke volume increase.

_____ 2. Total resistance to blood flow.

_____ 3. Effect on blood pressure if peripheral resistance increases.

_____ 4. Effect on blood pressure if arteries and arterioles constrict.

☞ Blood pressure is regulated by baroreceptor reflexes, chemoreceptor reflexes, and hormonal mechanisms.

Baroreceptor Reflexes

66The baroreceptor reflexes detect changes in blood pressure.99

Using the terms provided, complete the following statements:

Baroreceptors
Decrease
Increase

Pressure receptors that respond to stretch produced by blood pressure are called __(1)__. They are located in the carotid sinus and aortic arch. Action potentials from the baroreceptors pass to the medulla oblongata, which produces responses in blood vessels and in the heart. A sudden decrease in blood pressure is detected by these receptors and activates the baroreceptor reflex. As a result, the vasomotor center stimulates blood vessels to constrict, causing vasomotor tone to __(2)__. The change in blood vessel diameter causes peripheral resistance to __(3)__, and this in turn causes blood pressure to __(4)__. At the same time the cardioregulatory center causes heart rate to __(5)__, and this change in heart rate causes blood pressure to __(6)__.

1. _____

2. _____

3. _____

4. _____

5. _____

6. _____

 The baroreceptor reflexes are important in regulating blood pressure on a moment-to-moment basis.

Chemoreceptor Reflexes

66The chemoreceptor reflexes respond to changes in blood oxygen, carbon dioxide, and pH.99

Using the terms provided, complete the following statements:

Chemoreceptors
Decrease(s)
Increase(s)

Receptors that respond to oxygen, carbon dioxide, and pH are called __(1)__. They are located in the carotid bodies, aortic bodies, and medulla oblongata. A decrease in blood oxygen, an increase in carbon dioxide, or decrease in blood pH is detected by these receptors and activates the chemoreceptor reflex. As a result, vasomotor tone __(2)__, and blood vessels constrict. The change in blood vessel diameter causes peripheral resistance to __(3)__, and this in turn causes blood pressure to __(4)__. The change in blood pressure __(5)__ blood flow to the lungs, which helps to increase blood oxygen levels and decrease blood carbon dioxide levels.

1. _____

2. _____

3. _____

4. _____

5. _____

The chemoreceptor reflexes function under emergency conditions and usually do not play an important role in the regulation of the cardiovascular system.

192

Hormonal Mechanisms

"" *Hormonal mechanisms operate slower than nervous mechanisms.* **""**

Match these terms with the correct statement or definition:

Aldosterone   Epinephrine
Angiotensin   Renin

_____ 1. Released by the adrenal medulla; increases blood pressure by increasing heart rate and stroke volume.

_____ 2. Causes vasoconstriction of blood vessels in the skin and viscera, and vasodilation of blood vessels in skeletal and cardiac muscle.

_____ 3. Released by the kidneys in response to a decrease in blood pressure.

_____ 4. Activated by renin and other enzymes; increases blood pressure by causing vasoconstriction.

_____ 5. Acts on the adrenal cortex to cause increased aldosterone secretion.

_____ 6. Increases sodium and water uptake in the kidneys; maintains or increases blood pressure by maintaining or increasing blood volume.

Long-term and Short-term Regulation of Blood Pressure

"" *Deviations in blood pressure are corrected rapidly or gradually.* **""**

Match these terms with the correct statement or definition:

Adrenal medullary mechanism
Baroreceptor reflex
Renin-angiotensin-aldosterone mechanism

_____ 1. These two are most important for short-term regulation; sensitive to sudden changes in blood pressure they respond quickly to return blood pressure to normal.

_____ 2. Most important for long-term regulation because it responds slowly to gradually return blood pressure to normal.

Fetal Circulation

Prior to birth blood circulates to the placenta and bypasses the lungs.

Match these terms with the correct statement or definition:

Ductus arteriosus   Umbilical arteries
Foramen ovale   Umbilical vein

_____ 1. Opening between the right and left atria; allows blood to bypass the lungs.

_____ 2. Connection between the pulmonary trunk and aorta; allows blood to bypass the lungs.

_____ 3. Carries blood to the placenta.

_____ 4. Contains oxygen- and nutrient-rich blood.

Hypertension and Circulatory Shock

High or low deviations from normal blood pressure can be dangerous.

Match these terms with the correct statement or definition:

Anaphylactic shock   Hypertension
Cardiogenic shock   Neurogenic shock
Circulatory shock   Plasma loss shock
Heart failure   Septic shock
Hemorrhagic shock

_____ 1. Elevated blood pressure; increases the likelihood that blood vessels rupture and thrombi or emboli form; increases the rate at which arteriosclerosis develops.

_____ 2. Progressive weakening of cardiac muscle.

_____ 3. General term for inadequate blood flow throughout the body.

_____ 4. Results from extensive bleeding.

_____ 5. Results from severe burns, dehydration, severe diarrhea, or severe vomiting.

_____ 6. Results in vasodilation in response to emotions or anesthesia.

_____ 7. Caused by an allergic reaction that produces vasodilation and increased capillary permeability.

_____ 8. Caused by an infection that produces vasodilation and increased capillary permeability.

_____ 9. Heart stops pumping as a result of heart attack or electrocution.

1. Name the three layers or tunics of a blood vessel.

2. List the types of blood vessels, starting at the heart, going to tissues, and returning to the heart.

3. List the steps in determining blood pressure using the auscultatory method.

4. Name two ways blood flow is regulated. Which way controls blood flow within tissues, and which way controls blood flow to different regions of the body?

5. State the effect of an increase in heart rate, stroke volume, and peripheral resistance on blood pressure.

6. List two nervous mechanisms for regulating blood pressure. Which one is the most important under normal conditions?

7. List two hormonal mechanisms for regulating blood pressure.

Place the letter corresponding to the correct answer in the space provided.

_____1. Given the following blood vessels:
1. arteriole
2. elastic artery
3. muscular artery

Choose the arrangement that lists the blood vessels in the order a red blood cell passes through them as the red blood cell leaves the heart and travels to a tissue.
a. 2, 3, 1
b. 2, 1, 3
c. 3, 1, 2
d. 3, 2, 1

_____2. Given the following blood vessels:
1. capillary
2. vein
3. venule

Choose the arrangement that lists the blood vessels in the order a red blood cell passes through them as the red blood cell leaves a tissue and returns to the heart.
a. 1, 2, 3
b. 1, 3, 2
c. 3, 1, 2
d. 3, 2, 1

_____3. Comparing and contrasting veins and arteries,
a. veins have thicker walls than arteries.
b. veins have a greater amount of smooth muscle than arteries.
c. veins have a tunica media and arteries do not.
d. veins have valves and arteries do not.
e. all of the above

_____4. Given the following blood vessels:
1. aorta
2. inferior vena cava
3. pulmonary arteries
4. pulmonary veins

Which of the vessels carries oxygen rich blood?
a. 1, 3
b. 1, 4
c. 2, 3
d. 2, 4

_____5. Given the following arteries:
1. brachiocephalic artery
2. left common carotid artery
3. left subclavian artery

Choose the arrangement that lists the arteries in the order they branch off the aortic arch.
a. 1, 2, 3
b. 1, 3, 2
c. 2, 1, 3
d. 2, 3, 1

_____6. Which of the following arteries is correctly matched with the part of the body supplied by the artery?
a. brachiocephalic artery - left upper limb
b. brachiocephalic artery - right side of the head
c. left common carotid - left upper limb
d. left subclavian - left side of the head

_____ 7. Given the following arteries:
1. common carotid
2. internal carotid
3. subclavian artery
4. vertebral artery

Choose the arrangement that lists the arteries in order going from the aorta to the Circle of Willis in the brain.
a. 1, 2
b. 2, 1
c. 3, 1
d. 4, 2

_____ 8. Given the following arteries:
1. common carotid
2. internal carotid
3. subclavian artery
4. vertebral artery

Choose the arrangement that lists the arteries in order going from the aorta to the Circle of Willis in the brain.
a. 3, 2
b. 3, 4
c. 4, 1
d. 4, 3

_____ 9. Given the following arteries:
1. axillary artery
2. brachial artery
3. subclavian artery

Choose the arrangement that lists the arteries in order going toward the right hand.
a. 1, 2, 3
b. 1, 3, 2
c. 3, 1, 2
d. 3, 2, 1

_____ 10. Artery most commonly used to take the pulse near the wrist?
a. basilar
b. brachial
c. cephalic
d. radial
e. ulnar

_____ 11. Which of the following blood vessels is a paired artery?
a. celiac artery
b. renal artery
c. inferior mesenteric
d. superior mesenteric

_____ 12. Which of the following arteries is correctly matched with the organ it supplies?
a. celiac artery - stomach
b. inferior mesenteric artery - small intestine
c. superior mesenteric artery - last part of the colon
d. suprarenal artery - kidney

_____ 13. Given the following arteries:
1. common iliac
2. external iliac
3. femoral
4. popliteal

Choose the arrangement that lists the arteries in order going from the aorta to the knee.
a. 1, 2, 3, 4
b. 1, 2, 4, 3
c. 2, 1, 3, 4
d. 2, 1, 4, 3

_____ 14. Given the following veins:
1. brachiocephalic
2. internal jugular
3. superior vena cava

Choose the arrangement that lists the veins in the correct order going from the brain to the heart.
a. 1, 2, 3
b. 1, 3, 2
c. 2, 1, 3
d. 2, 3, 1

_____ 15. Given the following veins:
1. axillary
2. basilic
3. brachial
4. cephalic

Which of the veins are superficial veins?
a. 1, 3
b. 1, 4
c. 2, 3
d. 2, 4

_____16. Which vein is correctly matched with
its function?
a. azygous vein - receives blood from
the thoracic wall
b. brachiocephalic vein - receives
blood from the head and upper limbs
c. median cubital vein - common site for
drawing blood
d. all of the above

_____17. Which of the following organs drains
directly into the inferior vena cava?
a. kidney
b. stomach
c. small intestine
d. spleen

_____18. The hepatic portal vein
a. begins with capillaries in the
digestive system.
b. carries blood to the liver.
c. carries blood rich in nutrients.
d. all of the above

_____19. The external iliac vein
a. drains blood from the pelvis.
b. joins the inferior vena cava.
c. receives blood from the femoral vein.
d. all of the above

_____20. The great saphenous vein
a. empties into the femoral vein.
b. is a superficial vein of the lower
limb.
c. is often used for coronary bypass
surgery.
d. all of the above

_____21. The sound a person hears when
measuring blood pressure using the
auscultatory method is
a. caused by the closing of the AV
valves.
b. caused by the closing of the
semilunar valves.
c. caused by turbulent blood flow in the
arteries.
d. the pulse.

_____22. Concerning fluid movement at the tissue
level,
a. the amount of fluid leaving the
arteriolar end of a capillary is equal
to the amount of fluid entering the
venous end.
b. fluid enters tissues through
lymphatic capillaries.
c. edema results if more fluid moves
into a tissue than moves out.
d. all of the above

_____23. Blood flow through a tissue
a. regulated by local control results
from relaxation and contraction of
precapillary sphincters.
b. decreases in response to a decrease in
blood oxygen.
c. decreases in response to an increase in
blood carbon dioxide.
d. all of the above

_____24. An increase in blood pressure can result
from an increase in
a. peripheral resistance.
b. heart rate.
c. stroke volume.
d. all of the above

_____25. If blood pressure increases, the expected
response through the baroreceptor
reflex is
a. increased sympathetic nervous
system activity.
b. a decrease in peripheral resistance.
c. increased vasomotor center activity.
d. increased heart rate.

_____26. A sudden release of epinephrine from
the adrenal medulla would
a. decrease heart rate.
b. increase urine production.
c. cause vasoconstriction of visceral
blood vessels.
d. cause vasoconstriction of blood
vessels in skeletal muscle.

_____27. In response to an increase in blood
pressure, the secretion or production of
which of the following would decrease?
a. aldosterone
b. angiotensin
c. renin
d. all of the above

Use a separate sheet of paper to complete this section.

1. For each of the following destinations, name all the blood vessels a red blood cell would pass through if it started its journey in the left ventricle and returned to the right atrium.
   A. Liver
   B. Small intestine
   C. Urinary bladder
   D. Skin of the right lateral forearm.
   E. Skin of the posterior knee.

2. When a thrombus blocks a coronary artery, coronary angioplasty can be performed to unplug the artery. In this procedure, a tube is passed through blood vessels to the site of the blockage. The end of the tube has a balloon that expands to breakup the blockage. Would you recommend starting this procedure at the femoral artery or the femoral vein? Explain.

3. It is discovered that a patient has colon cancer. The doctor immediately orders a liver scan to test for liver cancer. Explain why this makes sense.

4. Ima Fan loves to go to movies. After sitting in a movie for several hours she often develops edema in her legs and feet. Explain how this occurs.

5. After a long and leisurely lunch at a restaurant, sometimes elderly people faint when they stand up to leave the restaurant. Explain how this happens (Hint: assume that a homeostatic mechanism is not working as well as when they were younger).

6. An adult grasps a child by her wrists and swings her around and around. What should happen to the child's heart rate. Give two reasons for this change.

7. Buster Hart has a myocardial infarct (heart attack) and his blood pressure drops. Explain why his blood pressure drops and describe the neural mechanisms that would attempt to compensate. In Buster's case blood pressure was abnormally low for a few days following the myocardial infarct. Gradually, however, it returned to normal. Explain how this happened.

ANSWERS TO CHAPTER 13

General Features of Blood Vessel Structure
- A. 1. Tunica intima; 2. Tunica media;
  3. Tunica externa
- B. 1. Artery; 2. Elastic arteries; 3. Muscular arteries;
  4. Arterioles; 5. Capillary; 6. Precapillary
  sphincters; 7. Vein; 8. Venules; 9. Valves

Understanding Blood Vessels
1. Arteriosclerosis; 2. Atherosclerosis; 3. Varicose
veins; 4. Phlebitis; 5. Gangrene

Pulmonary Circulation
1. Pulmonary trunk; 2. Pulmonary arteries;
3. Pulmonary veins

Systemic Circulation: Arteries
- A. 1. Ascending aorta; 2. Aortic arch; 3. Descending
  aorta; 4. Thoracic aorta
- B. 1. Right common carotid artery; 2. Right
  subclavian artery; 3. Brachiocephalic artery;
  4. Left common carotid artery; 5. Left subclavian
  artery
- C. 1. Axillary artery; 2. Brachial artery; 3. Radial
  artery; 4. Ulnar artery
- D. 1. Celiac artery; 2. Suprarenal artery; 3. Renal
  artery; 4. Superior mesenteric artery; 5. Common
  iliac artery; 6. Testicular artery; 7. Inferior
  mesenteric artery
- E. 1. External iliac artery; 2. Femoral artery;
  3. Popliteal artery; 4. Internal iliac artery

Systemic Circulation: Veins
- A. 1. Subclavian vein; 2. Brachiocephalic vein;
  3. Superior vena cava; 4. Inferior vena cava;
  5. Pulmonary veins; 6. Internal jugular vein;
  7. External jugular vein
- B. 1. Cephalic vein; 2. Basilic vein; 3. Median cubital
  vein; 4. Brachial vein; 5. Axillary vein
- C. 1. Hepatic veins; 2. Testicular vein; 3. External
  iliac vein; 4. Internal iliac vein; 5. Common iliac
  vein; 6. Inferior vena cava; 7. Renal vein;
  8. Suprarenal vein

- D. 1. Inferior vena cava; 2. Hepatic veins; 3. Hepatic
  portal vein; 4. Superior mesenteric vein;
  5. Inferior mesenteric vein; 6. Splenic vein;
  7. Gastric vein
- E. 1. Great saphenous vein; 2. Small saphenous
  vein; 3. Popliteal vein; 4. Femoral vein

The Physiology of Circulation
- A. 1. Blood pressure; 2. Systolic pressure;
  3. Diastolic pressure; 4. Auscultatory; 5. Systolic
  pressure; 6. Diastolic pressure; 7. Pulse
- B. 1. Diffusion; 2. Out of capillary; 3. Into capillary;
  4. Lymphatic capillaries; 5. Edema

Local Control of Blood Vessels
1. Local control; 2. Relaxation; 3. Nervous control;
4. Vasomotor center; 5. Vasomotor tone;
6. Nervous control; 7. Local control

Regulation of Arterial Pressure
1. Increases; 2. Peripheral resistance; 3. Increases;
4. Increases

Baroreceptor Reflexes
1. Baroreceptors; 2. Increase; 3. Increase; 4. Increase;
5. Increase; 6. Increase

Chemoreceptor Reflexes
1. Chemoreceptors; 2. Increases; 3. Increase;
4. Increase; 5. Increases

Hormonal Mechanisms
1. Epinephrine; 2. Epinephrine; 3. Renin;
4. Angiotensin; 5. Angiotensin; 6. Aldosterone

Long-term and Short-term Regulation of Blood Pressure
1. Baroreceptor reflex and adrenal medullary
mechanism; 2. Renin-angiotensin-aldosterone
mechanism

Fetal Circulation
1. Foramen ovale; 2. Ductus arteriosus; 3. Umbilical
arteries; 4. Umbilical vein

Hypertension and Circulatory Shock
1. Hypertension; 2. Heart failure; 3. Circulatory
shock; 4. Hemorrhagic shock; 5. Plasma loss shock;
6. Neurogenic shock; 7. Anaphylactic shock; 8. Septic
shock; 9. Cardiogenic shock

1. Tunica intima, tunica media, and tunica externa
2. Elastic arteries, muscular arteries, arterioles,
   capillaries, venules, veins
3. A pressure cuff is used to collapse the brachial
   artery. Pressure is released and a stethoscope is
   used to detect turbulent blood flow. Systolic
   pressure occurs when sound is first heard and
   diastolic pressure when the sound disappears
4. Local control (precapillary sphincters) and nervous
   control (arterioles) regulate blood flow within tissues;
   nervous control (muscular arteries) also regulates
   blood flow to different regions of the body
5. All increase blood pressure

6. Baroreceptor (most important) and chemoreceptor reflexes (emergency conditions)

7. Epinephrine and renin-angiotensin-aldosterone

1. A. The red blood cell passes through an elastic artery, muscular artery and arteriole.

2. B. The red blood cell passes through a capillary, venule, and vein.

3. D. Only veins have valves. Veins are thinner walled with less smooth muscle than arteries. Both veins and arteries have all three tunics.

4. B. Oxygen rich blood returns from the lungs to the heart through the pulmonary veins, passes through the left side of the heart, and exits through the aorta.

5. A. The correct order is the brachiocephalic, left common carotid, and left subclavian.

6. B. The brachiocephalic artery branches to form the right common carotid, which supplies the right side of the head, and the right subclavian, which supplies the right upper limb. The left common carotid supplies the left side of the head, and the left subclavian supplies the left upper limb.

7. A. The common carotid artery branches to form the internal and external carotid arteries. The internal carotid goes to the brain, whereas the external carotid supplies the external head.

8. B. The subclavian arteries give rise to the vertebral arteries, which supply the brain.

9. C. The subclavian artery becomes the axillary artery, which then becomes the brachial artery.

10. D. The radial artery passes over the radius near the wrist, and the pulse can be easily taken by slight compression of the artery against the bone.

11. B. The major paired arteries are the renal, suprarenal, and testicular or ovarian arteries. The unpaired arteries are the celiac, inferior mesenteric, and superior mesenteric arteries.

12. A. The celiac artery supplies the stomach, liver, and spleen.

13. A. The order is common iliac, external iliac, femoral, and popliteal arteries.

14. C. The internal jugular drains the brain. It joins with the subclavian vein to form the brachiocephalic vein, which empties into the superior vena cava.

15. D. The basilic and cephalic are superficial veins of the upper limb. The axillary and brachial are deep veins.

16. D. All of the vessels are correctly matched.

17. A. The kidneys empty into the inferior vena cava. The stomach, small intestine, and spleen empty into the hepatic portal vein.

18. D. The hepatic portal vein begins with capillaries in the digestive tract and ends with capillaries in the liver. It functions to carry nutrient rich blood from the digestive tract to the liver, in which the nutrients are processed.

19. C. The external iliac vein receives blood from the femoral vein, which drains the lower limb. The external iliac vein joins the common iliac vein, which then joins the inferior vena cava. The internal iliac vein drains blood from the pelvis.

20. D. All the choices are true.

21. C. Blood flows in the brachial artery, where blood pressure is determined by the ausculatory method. The pressure cuff constricts the brachial artery, resulting in turbulent flow, which produces the sounds. The heart sounds caused by closure of the AV and semilunar valves are not heard. The pulse is a pressure wave that can be felt.

22. C. Edema is a buildup of fluid in tissue resulting from more fluid movement into the tissue than moves out. The amount of fluid leaving the arterial ends of capillaries is greater than the amount returning in the venous ends. Consequently, more fluid enters tissues than leaves. The excess fluid leaves the tissue through lymphatic capillaries.

23. A. Local control of blood flow results from the relaxation and contraction of the precapillary sphincters. The precapillary sphincters relax, increasing blood flow, in response to a decrease in blood oxygen or an increase in blood carbon dioxide.

24. D. Blood pressure increases when heart rate, stoke volume, or peripheral resistance increase.

25. B. If blood pressure increased, the baroreceptor reflex would cause a decrease in peripheral resistance by causing vasodilation. This response is mediated by decreased sympathetic activity through the vasomotor center.

26. **C.** Epinephrine causes vasoconstriction of visceral blood vessels and vasodilation of skeletal muscle blood vessels. This routes blood from the viscera to skeletal muscle. Epinephrine also increases heart rate and stroke volume.

27. **D.** The increase in blood pressure results in less renin secretion by the kidneys. Consequently, there is less conversion of angiotensinogen and less production of angiotensin. Lower levels of angiotensin result in less stimulation of the adrenal cortex and less aldosterone production.

 ## FINAL CHALLENGES

1. a. Aorta, celiac artery, liver, hepatic veins, inferior vena cava
    b. Aorta, superior mesenteric artery, small intestine, superior mesenteric vein, hepatic portal vein, liver, hepatic vein, inferior vena cava
    c. Aorta, common iliac artery, internal iliac artery, urinary bladder, internal iliac vein, common iliac vein, inferior vena cava
    d. Aorta, brachiocephalic artery, subclavian artery, axillary artery, brachial artery, radial artery, skin, cephalic vein, axillary vein, subclavian vein, brachiocephalic vein, superior vena cava
    e. Aorta, common iliac artery, external iliac artery, femoral artery, popliteal artery, skin, great saphenous vein, femoral vein, external iliac vein, common iliac vein, inferior vena cava

2. Start at the femoral artery. The tube moves through the following vessels: femoral artery, external iliac artery, common iliac artery, aorta, coronary artery.

3. The hepatic portal vein drains blood from the colon and carries it to the liver. Cancerous cells from the colon tumor could escape from the colon, become lodged within the capillaries of the liver, and start a liver tumor.

4. Normally compression of veins by skeletal muscle helps to move blood from the veins of the lower limbs back toward the heart, and the valves of the veins prevent backflow of the blood. By sitting for an extended period without this muscular activity, the blood tends to accumulate in the lower limbs. Fluid from the accumulated blood moves into the tissues and causes edema.

5. Sitting has resulted in a shift of fluid from the elderly person's blood into the tissues of the lower limbs (see Final Challenge Question #4). This causes a reduction in blood volume and blood pressure.

Also, as a result of the demands of digestion, blood is rerouted to the digestive organs. Consequently there is reduced blood flow toward the brain. When the elderly person stands up, there is an even greater tendency for blood to remain in the lower limbs (due to gravity). Normally, the baroreceptor reflex corrects for the drop in blood pressure by increasing heart rate and stroke volume. However, if this mechanism does not respond rapidly enough, blood pressure does not increase, blood delivery to the brain is impaired, and the person faints.

6. The child's heart rate should increase. The excitement of being swung around and around can cause release of epinephrine from the adrenal medulla. The epinephrine causes heart rate to increase. Swinging the child also causes blood to move toward the child's feet. This reduces blood return to the heart. Venous return decreases and cardiac output decreases (Starling's law). Consequently blood pressure decreases. The baroreceptor reflex detects the drop in blood pressure and compensates by increasing heart rate and stroke volume.

7. The myocardial infarct damages the heart and reduces its ability to pump blood. With a decrease in cardiac output there is a decrease in blood pressure and blood oxygen. The baroreceptor reflexes and the chemoreceptor reflexes compensate by increasing heart rate, stroke volume, and vasoconstriction (peripheral resistance), all of which increase blood pressure. In the days following the myocardial infarct long-term mechanisms compensate for the still low blood pressure. The renin-angiotensin-aldosterone mechanism acts to increase vasoconstriction and reduce fluid loss in the kidneys. With an increase in blood volume, blood pressure increases.

The Lymphatic System and Immunity

FOCUS: The lymphatic system includes lymph, lymphocytes, lymph vessels, lymph nodes, tonsils, the spleen, and the thymus gland. The lymphatic system helps maintain fluid balance in the tissues, absorbs fats and other substances from the digestive tract, and is part of the body's defense system.

CONTENT LEARNING ACTIVITY

Lymph Vessels

" *The lymphatic system, unlike the circulatory system, only carries fluid away from the tissues.* **"**

Match these terms with the correct statement or definition:

Lymph capillaries   Right lymphatic duct
Lymph vessels   Thoracic duct

_____ 1. Tiny, closed-ended vessels consisting of simple squamous epithelium.

_____ 2. One-way valves present; lymph capillaries join to form these.

_____ 3. Large lymph vessel that empties into the right subclavian vein; drains the right head, neck, chest, and upper limb.

_____ 4. Large lymph vessel that empties into the left subclavian vein; drains all but the right head, neck, chest, and upper limb.

 Contraction of surrounding skeletal muscle, contraction of smooth muscle in lymph vessel walls, and pressure changes in the thorax during respiration are factors that assist the movement of lymph through lymph vessels.

Lymphatic Organs

""Lymphatic organs contain lymphatic tissue, which consists of many lymphocytes and other cells.**""**

Using the terms provided, complete the following statements:

Divide   Microorganisms
Immune system   Red bone marrow
Lymphatic organs

The lymphocytes originate from _(1)_ and are carried by the blood to _(2)_. When the body is exposed to _(3)_ or foreign substances, the lymphocytes _(4)_ and increase in number. Lymphocytes are part of the _(5)_ response that destroys microorganisms and foreign substances.

1. _____

2. _____

3. _____

4. _____

5. _____

Tonsils

""There are three groups of tonsils.**""**

A. Match these terms with the correct statement or definition:

Lingual tonsil   Pharyngeal tonsil
Palatine tonsils

_____

_____

_____

1. "The tonsils" located on each side of the posterior opening of the oral cavity.

2. Located near the internal opening of the nasal cavity; called the adenoid when enlarged.

3. Located on the posterior surface of the tongue.

B. Match these terms with the correct parts labeled in Figure 14-1:

Lingual tonsil
Palatine tonsil
Pharyngeal tonsil

1. _____

2. _____

3. _____

Lymph Nodes

66_Lymph nodes are small, round structures distributed along the various lymph vessels._**99**

Using the terms provided, complete the following statements:

Afferent   Immune system
Efferent   Lymphocytes
Germinal centers   Macrophages

Lymph enters the lymph node through _(1)_ vessels, passes through the lymphatic tissue, and exits through _(2)_ vessels. As lymph moves through lymph nodes, two things happen. One function is activation of the _(3)_. Microorganisms in the lymph can stimulate _(4)_ in the lymphatic tissue to start dividing. These areas of rapidly dividing lymphocytes are called _(5)_. Another function of lymph nodes is the removal (phagocytosis) of microorganisms and other foreign substances by _(6)_.

1. _____
2. _____
3. _____
4. _____
5. _____
6. _____

Spleen

66_The spleen is roughly the size of a clenched fist._**99**

Using the terms provided, complete the following statements:

Blood   Red blood cells
Foreign substances   Reservoir
Lymphocytes   Splenectomy
Macrophages

The spleen filters _(1)_ instead of lymph. _(2)_ in the spleen can be stimulated; therefore the immune system is activated in response to microorganisms and _(3)_. _(4)_ in the spleen remove microorganisms, foreign substances, and worn out _(5)_ by phagocytosis. The spleen also functions as a blood _(6)_, holding a small volume of blood. If the spleen is injured, a _(7)_, removal of the spleen, is performed to stop the bleeding.

1. _____
2. _____
3. _____
4. _____
5. _____
6. _____
7. _____

Thymus

"*The thymus is a bilobed gland roughly triangular in shape.***"**

Using the terms provided, complete the following statements:

Lymphatic tissue   Mediastinum
Lymphocytes   Microorganisms
Maturation   Puberty

The thymus is located in the __(1)__, the partition dividing the thoracic cavity into right and left parts. In a newborn, the thymus is large, and continues to grow until __(2)__, after which it decreases in size. Large numbers of __(3)__ are produced in the thymus, but for unknown reasons, most degenerate. The thymus functions as a site for processing and __(4)__ of lymphocytes. While in the thymus, lymphocytes do not respond to foreign substances, but after they have matured, thymic lymphocytes travel to other __(5)__, where they help protect against __(6)__ and other foreign substances.

1. _____

2. _____

3. _____

4. _____

5. _____

6. _____

Disorders of the Lymphatic System

"*Many infectious diseases produce symptoms associated with the lymphatic system.***"**

Match these terms with the correct statement or definition:

Bubonic plague   Lymphangitis
Elephantiasis   Lymphoma
Lymphadenitis

_____ 1. General term for inflammation of the lymph nodes; causes them to become enlarged and tender.

_____ 2. Inflammation of the lymph vessels.

_____ 3. A neoplasm (tumor) of lymphatic tissue; e. g., Hodgkin's disease.

_____ 4. Caused by bacteria that are transferred to humans from rats by the bite of a flea; often causes inguinal lymph nodes to swell.

_____ 5. Caused by roundworms that block lymph vessels; offspring of adult worms are passed to other humans by mosquitoes.

Immunity

66 *Immunity is the ability to resist damage from foreign substances such as microorganisms and* 99 *harmful chemicals.*

Match these terms with the correct statement or definition:

Nonspecific resistance
Specific resistance

_____ 1. Identical response each time the body is exposed to a given substance.

_____ 2. Response to a substance during the second exposure is faster and stronger than during the first exposure.

_____ 3. Response results in immunity to harmful effects of a bacteria or virus.

Nonspecific Resistance

66 *Nonspecific resistance includes mechanical mechanisms, chemical mediators, and the* 99 *inflammatory response.*

A. Match these terms with the correct statement or definition:

Complement   Lysozyme
Histamine   Skin and mucous membranes
Interferons

_____ 1. Barriers that prevent entry of microorganisms.

_____ 2. Chemical substance in tears and saliva that kills microorganisms.

_____ 3. Chemical that promotes inflammation by causing vasodilation and increased vascular permeability.

_____ 4. A group of at least 11 proteins found in plasma; once activated they promote inflammation and phagocytosis, and can lyse bacterial cells.

_____ 5. Proteins that protect the body against viral infections.

B. Match these terms with the correct statement or definition:

Basophils   Mast cells
Eosinophils   Neutrophils
Macrophages   Pus

_____ 1. Small phagocytic cells that usually are the first cells to enter infected tissues from the blood.

_____ 2. Primarily an accumulation of dead neutrophils at the site of an infection.

_____ 3. Monocytes that leave the blood, enter tissues, and enlarge fivefold.

_____ 4. WBCs that release inflammatory chemicals such as histamine.

_____ 5. Cells found in connective tissue that release inflammatory chemicals.

_____ 6. WBCs that release chemicals that break down inflammatory chemicals.

White blood cells are produced in red bone marrow and lymphatic tissue and are released into the blood. Chemicals such as complement and histamine, released by microorganisms or damaged tissues, attract WBCs.

Inflammatory Response

66*The inflammatory response is a complex sequence of events involving both chemicals and cells.***99**

Using the terms provided, complete the following statements:

Chemicals   Phagocytes
Complement   Vascular permeability
Fibrin   Vasodilation

Most inflammatory responses are similar. The microorganism itself or damage to tissues causes the release or activation of _(1)_ such as histamine, prostaglandins, complement, and others. The chemicals produce several effects: _(2)_ increases blood flow and brings phagocytes and other leukocytes to the area; attraction of _(3)_, which leave the blood and enter tissue; and increased _(4)_ allowing fibrin and complement to enter the tissue from the blood. _(5)_ prevents the spread of infection by walling off the infected area. _(6)_ further enhances the inflammatory response and attracts additional phagocytes.

1. _____

2. _____

3. _____

4. _____

5. _____

6. _____

Specific Resistance

66*Specific resistance involves the ability to recognize, respond to, and remember a particular substance.***99**

A. Match these terms with the correct statement or definition:

Antigens   Self antigens
Foreign antigens

_____ 1. General term for substances that stimulate specific resistance responses.

_____ 2. Antigens introduced from outside the body.

_____ 3. Molecules produced by the body that stimulate an immune system response, e.g., molecules produced by tumor cells.

B. Match these terms with the correct statement or definition:

Antibody-mediated immunity   Helper T cells
B cells   T cells
Cell-mediated immunity

_____ 1. Lymphocytes that give rise to cells which produce antibodies.

_____ 2. Immunity produced by antibodies in plasma; humoral immunity.

_____ 3. Lymphocytes responsible for cell-mediated immunity.

_____ 4. T cells that regulate cell-mediated and antibody-mediated immunity by stimulating B cells or other T cells.

Autoimmune disease results when self antigens stimulate destruction of normal tissue, e.g., rheumatoid arthritis.

Antibody-Mediated Immunity

66 *Antibody-mediated immunity is effective against extracellular antigens, and is also involved with* **99** *allergic reactions.*

Match these terms with the correct statement or definition:

Memory B cells
Plasma cells

Primary response
Secondary (memory) response

_____   1. Result of first exposure of B cell to an antigen.

_____   2. Cells that produce antibodies.

_____   3. When exposed to an antigen, these cells rapidly divide to produce plasma cells.

_____   4. Response that occurs when the immune system is exposed to an antigen against which it has already produced a primary response.

Understanding Antibodies

66 *Large amounts of antibodies are in plasma.* **99**

Using the terms provided, complete the following statements:

Antigen
Diseases
Immunoglobulins
Inflammatory response

Monoclonal antibodies
Tumor cells
Y-shaped

1. _____

2. _____

3. _____

4. _____

5. _____

6. _____

7. _____

Antibodies, also called gamma globulins or _(1)_, are proteins produced in response to a(n) _(2)_. They are _(3)_ molecules; the end of each "arm" attaches to the antigen. This part of a particular antibody is specific for a particular antigen. Binding of antibodies to antigens can stimulate a(n) _(4)_, activating chemicals and cells and eventually destroying the antigens. _(5)_ are a pure antibody preparation that is specific for only one antigen. They are called monoclonal because only one kind of antibody is produced from a clone of B cells. The clone of B cells is fused with _(6)_ to obtain a rapidly dividing group of cells that produce only one type of antibody. Monoclonal antibodies are used to test for pregnancy, diagnose _(7)_, and treat cancer.

Cell-Mediated Immunity

66 *Cell-mediated immunity is a function of T cells and is most effective against microorganisms that* **99**
live inside the cells of the body.

Match these terms with the
correct statement or definition:

Cytotoxic T cells
T memory cells

_____ 1. Cells responsible for the cell-mediated immunity response; produce
lymphokines and lyse virus-infected cells, tumor cells, or tissue
transplant cells.

_____ 2. T cells that provide a secondary response and long-lasting immunity.

☞ Cytotoxic T cells have two main effects. First, they release chemicals that promote
inflammation and phagocytosis, and, they come into contact with other cells and cause them to
lyse (rupture).

Acquired Immunity

66 *There are four ways to acquire immunity: active natural, active artificial, passive natural, and* **99**
passive artificial.

Match these terms with the
correct statement or definition:

Active artificial immunity   Passive artificial immunity
Active natural immunity   Passive natural immunity
Antiserum   Vaccine

_____ 1. Results from natural exposure to an antigen that causes the body's
immune system to respond against the antigen.

_____ 2. Results when an antigen is deliberately introduced into an individual
to stimulate his immune system; also called vaccination.

_____ 3. Introduced antigen that produces active artificial immunity; usually
consists of a live, altered microorganism, dead microorganism, or part
of a microorganism.

_____ 4. Results from the transfer of antibodies from a mother to her child
across the placenta.

_____ 5. Results when antibodies are removed from a human or another
animal and injected into an individual.

_____ 6. General term used for antibodies that provide passive artificial
immunity.

Immune System Problems of Clinical Significance

66The immune system is involved in many conditions of clinical significance.99

A. Match these terms with the
correct statement or definition:

Allergen   Delayed hypersensitivity
Anaphylaxis   Immediate hypersensitivity
Asthma   Urticaria

_____ 1. Antigen that stimulates an allergic reaction.

_____ 2. Reaction produced within minutes of exposure to an allergen, (e. g., hay fever); caused by antibodies.

_____ 3. Allergic reaction in which mast cells or basophils release inflammatory chemicals in the lungs, causing constriction of airways.

_____ 4. Skin rash or localized swelling usually caused by an ingested allergen.

_____ 5. Systemic allergic reaction resulting in systemic vasodilation, a drop in blood pressure, and possibly death.

_____ 6. Reaction that takes hours to days to develop, (e. g., poison ivy); caused by T cells.

B. Match these terms with the
correct statement or definition:

AIDS   Immunodeficiency
Autoimmune disease   Immune surveillance
Human lymphocyte antigens Transplantation

_____ 1. Condition in which the immune system incorrectly treats self antigens as foreign antigens.

_____ 2. Failure of some part of the immune system to function properly, e. g., SCID, in which both B cells and T cells fail to form.

_____ 3. Caused by the human immunodeficiency virus (HIV); results in death of T cells.

_____ 4. Detection and destruction of tumor cells by the immune system.

_____ 5. Antigens on the surface of human cells that allow the immune system to distinguish between self cells and foreign cells.

_____ 6. Using donor tissue from another individual to replace damaged tissue; immune suppression of the patient's immune system is usually required.

1. List three basic functions of the lymphatic system.

2. List three factors that compress lymph vessels and move lymph toward the circulatory system.

3. List the functions performed by the lymph nodes, spleen, and thymus.

4. List four types of nonspecific resistance, and give an example of each.

5. List the two cell types responsible for most of the phagocytosis in the body.

6. Name the cell type that is responsible for antibody-mediated immunity and the cell type that is responsible for cell-mediated immunity.

7. Define a primary response and a memory response.

8. Give the two basic ways that cell-mediated immunity acts against an antigen.

Place the letter corresponding to the correct answer in the space provided.

_____ 1. The lymphatic system
   a. removes excess fluid from tissues.
   b. absorbs fats from the digestive tract.
   c. defends the body against microorganisms and other foreign substances.
   d. all of the above

_____ 2. Lymph vessels
   a. do not have valves.
   b. pass through at least one lymph node.
   c. from the right upper limb join the thoracic duct.
   d. all of the above

_____ 3. Lymph is moved through lymph vessels because of
   a. blood pressure.
   b. pressure changes in the digestive tract.
   c. contraction of surrounding skeletal muscle.
   d. all of the above

_____ 4. The tonsils
   a. provide protection against harmful material entering the nose and mouth.
   b. increase in size in adults.
   c. consist of four groups.
   d. all of the above

_____ 5. Lymph nodes
   a. produce lymphocytes.
   b. remove microorganisms and foreign substances from lymph.
   c. contain macrophages.
   d. all of the above

_____ 6. The spleen
   a. detects and responds to foreign substances in the blood.
   b. serves as a reservoir for lymph.
   c. produces new red blood cells.
   d. all of the above

_____ 7. The thymus
   a. increases in size in adults.
   b. produces lymphocytes that move to other lymph tissue.
   c. responds to foreign substances in the blood.
   d. all of the above.

_____ 8. A group of chemicals that is activated by a series of reactions, in which each component of the series activates the next component, and the activated chemicals promote inflammation and phagocytosis, is called
   a. histamine.
   b. antibodies.
   c. complement.
   d. interferons.
   e. lysozyme.

_____ 9. A substance produced by cells in response to infection by viruses and prevents viral replication in other cells is
   a. interferon.
   b. complement.
   c. fibrin.
   d. antibodies.
   e. histamine.

_____ 10. Neutrophils
   a. enlarge to become macrophages.
   b. account for most of the dead cells in pus.
   c. are usually the last cell type to enter infected tissues.
   d. are involved in the cleanup work in the late stages of an infection.
   e. all of the above

_____ 11. In addition to leukocytes, some cells found in connective tissue release inflammatory chemicals. These cells are called
   a. macrophages.
   b. eosinophils.
   c. neutrophils.
   d. mast cells.
   e. monocytes.

_____12. During the inflammatory response
   a. histamine and other chemical mediators are released.
   b. attraction of phagocytes occurs.
   c. fibrin enters tissue from the blood.
   d. blood vessels vasodilate.
   e. all of the above.

_____13. Antigens
   a. are foreign substances introduced into the body.
   b. are molecules produced by the body.
   c. stimulate a specific immune system response.
   d. all of the above

_____14. B cells
   a. produce plasma cells.
   b. are responsible for immediate hypersensitivity reactions.
   c. produce memory cells.
   d. all of the above

_____15. Helper T cells
   a. stimulate the activity of B cells.
   b. stimulate the activity of T cells.
   c. produce antibodies.
   d. both a and b

_____16. Antibodies
   a. are specific for one particular antigen.
   b. cannot stimulate the inflammatory response.
   c. are carbohydrates produced in response to an antigen.
   d. all of the above

_____17. Specific resistance involves
   a. both B cells and T cells.
   b. the production of memory cells.
   c. the ability to remember and respond to specific antigens.
   d. all of the above

_____18. The secondary antibody response
   a. is slower than the primary response.
   b. produces less antibodies than the primary response.
   c. prevents the appearance of disease symptoms.
   d. a and b

_____19. The type of lymphocyte responsible for the secondary antibody response?
   a. memory cell
   b. plasma cell
   c. cytotoxic T cell
   d. T helper cell
   e. mast cell

_____20. The type of immunity produced by vaccination would be
   a. active natural immunity.
   b. passive natural immunity.
   c. active artificial immunity.
   d. passive artificial immunity.

FINAL CHALLENGES

Use a separate sheet of paper to complete this section.

1. The central nervous system and bone marrow do not have lymph vessels. Why doesn't edema occur in these tissues?

2. Acquired immune deficiency syndrome (AIDS) is caused by the human immunodeficiency virus (HIV). If HIV infects mainly T cells, why are both antibody-mediated immunity and cell-mediated immunity affected?

3. Ivy Mann developed a poison ivy rash after a camping trip. His doctor prescribed a cortisol ointment to relieve the inflammation. A few weeks later Ivy scraped his elbow, which became inflamed. Because he had some of the cortisol ointment left over, he applied it to the scrape. Why or why not was the ointment a good idea for the poison ivy and for the scrape?

4. A young lady has just had her ears pierced. To her dismay, she finds that, when she wears inexpensive (but tasteful) jewelry, by the end of the day there is an inflammatory (allergic) reaction to the metal in the jewelry. Is this caused by antibody-mediated immunity or cell-mediated immunity?

5. A patient is taking a drug to prevent graft rejection. Would you expect the patient to be more or less likely to develop cancer? Explain.

ANSWERS TO CHAPTER 14

CONTENT LEARNING ACTIVITY

Lymph Vessels
1. Lymph capillaries; 2. lymph vessels; 3. Right lymphatic duct; 4. Thoracic duct

Lymphatic Organs
1. Red bone marrow; 2. Lymphatic organs; 3. Microorganisms; 4. Divide; 5. Immune system

Tonsils
A. 1. Palatine tonsils; Pharyngeal tonsil; Lingual tonsil
B. 1. Pharyngeal tonsil; 2. Palatine tonsil; 3. Lingual tonsil

Lymph nodes
1. Afferent; 2. Efferent; 3. Immune system; 4. Lymphocytes; 5. Germinal centers; 6. Macrophages

Spleen
1. Blood; 2. Lymphocytes; 3. Foreign substances; 4. Macrophages; 5. Red blood cells; 6. Reservoir; 7. Splenectomy

Thymus
1. Mediastinum; 2. Puberty; 3. Lymphocytes; 4. Maturation; 5. Lymphatic tissue; 6. Microorganisms

Disorders of the Lymphatic System
1. Lymphadenitis; 2. Lymphangitis; 3. Lymphoma; 4. Bubonic plague; 5. Elephantiasis

Immunity
1. Nonspecific resistance; 2. Specific resistance; 3. Specific resistance

Nonspecific Resistance
A. 1. Skin and mucous membrane; 2. Lysozyme; 3. Histamine; 4. Complement; 5. Interferons

B. 1. Neutrophils; 2. Pus; 3. Macrophages; 4. Basophils; 5. Mast cells; 6. Eosinophils

Inflammatory Response
1. Chemicals; 2. Vasodilation; 3. Phagocytes; 4. Vascular permeability; 5. Fibrin; 6. Complement

Specific Resistance
A. 1. Antigens; 2. Foreign antigens; 3. Self antigens
B. 1. B cells; 2. Antibody-mediated immunity; 3. T cells; 4. Helper T cells

Antibody-Mediated Immunity
1. Primary response; 2. Plasma cells; 3. Memory B cells; 4. Secondary (memory) response

Understanding Antibodies
1. Immunoglobulins; 2. Antigen; 3. Y-shaped; 4. Inflammatory response; 5. Monoclonal antibodies; 6. Tumor cells; 7. Diseases

Cell-Mediated Immunity
1. Cytotoxic T cells; 2. T memory cells

Acquired Immunity
1. Active natural immunity; 2. Active artificial immunity; 3. Vaccine; 4. Passive natural immunity; 5. Passive artificial immunity; 6. Antiserum

Immune System Problems of Clinical Significance
A. 1. Allergen; 2. Immediate hypersensitivity; 3. Asthma; 4. Urticaria (hives); 5. Anaphylaxis; 6. Delayed hypersensitivity
B. 1. Autoimmune disease; 2. Immunodeficiency; 3. AIDS; 4. Immune surveillance; 5. Human lymphocyte antigens; 6. Transplantation

QUICK RECALL

1. Maintains fluid balance in the tissues, absorbs fats and other substances from the digestive tract, and defends against microorganisms and other foreign substances

2. Contraction of surrounding skeletal muscles, contraction of smooth muscle in lymph vessels, and pressure changes in the thorax during respiration

4. Mechanical mechanisms: skin and mucous membrane form barrier, tears, saliva, urine, cilia and mucus remove microorganisms; Chemicals: histamine, complement, interferon; Cells: neutrophils, macrophages, mast cells, basophils, and eosinophils; Inflammatory response: involves many types of cells and chemicals

5. Neutrophils and macrophages

6. Antibody-mediated immunity: B cells; Cell-mediated immunity: T cells

7. Primary response: first exposure to an antigen; secondary response: exposure to an antigen after the primary response

8. Release chemicals that promote inflammation and phagocytosis or directly cause cells to lyse

MASTERY LEARNING ACTIVITY

1. **D.** The lymphatic system is involved in tissue fluid balance, fat absorption, and defense against microorganisms and other foreign substances.

2. **B.** Most lymph vessels pass through a lymph node. Lymph vessels have valves that ensure one-way flow, and lymph vessels from the right upper limb, right thorax, and right side of the head and neck join the right lymphatic duct. Vessels from the rest of the body empty into the thoracic duct.

3. **C.** Lymph is moved by contraction of surrounding skeletal muscle, contraction of smooth muscle in lymph vessels, and increased thoracic pressure during respiration.

4. **A.** There are three groups of tonsils: palatine, pharyngeal, and lingual. The tonsils decrease in size in adults, and form a protective ring of lymphatic tissue between the mouth, nose, and pharynx that helps to prevent harmful material from entering the pharynx from the nose and mouth.

5. **D.** Lymph nodes contain macrophages which remove microorganisms and other foreign substances by phagocytosis. Lymphocytes are produced in germinal centers of the lymph node.

6. **A.** The spleen detects and responds to foreign substances in the blood, destroys worn out red blood cells, and acts as a reservoir for red blood cells.

7. **B.** The thymus produces lymphocytes that move to other lymphatic tissue. However, lymphocytes in the thymus do not respond to foreign substances. With increasing age in the adult, the thymus decreases in size.

8. **C.** The series of reactions produces activated complement, which promotes inflammation and phagocytosis, and can lyse bacterial cells.

9. **A.** Interferon does not protect the cell producing the interferon, but does bind to other cells and prevents viral replication.

10. **B.** Neutrophils account for most of the dead cells in pus. Neutrophils are phagocytic cells that are usually the first cell type to enter infected tissues. Macrophages appear in infected tissue after neutrophils and are responsible for the cleanup of dead neutrophils and other cellular debris.

11. **D.** Mast cells are cells found in connective tissue that release inflammatory chemicals. Like macrophages, mast cells are located where microorganisms are likely to enter the body.

12. **E.** Damage to tissue causes the release of histamine and other chemical mediators that cause vasodilation, attract phagocytes, and increase vascular permeability (allowing the entry of fibrin, which walls off infected areas).

13. **D.** Antigens are substances that stimulate a specific immune system response. They can be foreign substances (foreign antigens) or molecules produced by the body (self antigens).

14. **D.** When B cells are stimulated by an antigen, they divide and produce plasma cells and memory cells. The plasma cells produce antibodies. B cells and the antibodies they produce are responsible for immediate hypersensitivity reactions.

15. **D.** Helper T cells are a special kind of T cell responsible for regulating the immune system. Helper T cells stimulate the activity of both B cells and other T cells. B cells produce antibodies.

16. **A.** Antibodies are specific for one particular antigen. Antibodies are proteins produced in response to an antigen, and they can stimulate the inflammatory response.

17. **D.** Specific immunity involves both B cells (antibody-mediated-immunity) and T cells (cell-mediated immunity). Both B cells and T cells produce memory cells, which have the ability to remember and respond to specific antigens.

18. C. The secondary response is more rapid and produces more antibodies that the primary response. Thus the secondary response effectively destroys the antigen and prevents the appearance of disease symptoms.

19. A. During the primary response memory cells are formed. They are responsible for the secondary response.

20. C. Active artificial immunity occurs when an antigen is deliberately introduced into an individual to stimulate his immune system. The introduced antigen is called a vaccine. Active natural immunity occurs when an individual is naturally exposed to an antigen that causes the individual's immune system to respond. Passive natural immunity occurs when antibodies pass through the placenta from mother to child, and passive artificial immunity occurs when antibodies (in antiserum) from a human or another animal are injected into the individual requiring immunity.

 FINAL CHALLENGES

1. Because the brain, spinal cord, and bone marrow are encased by bone, the bone prevents the tissues from swelling.

2. Although certain types of T cells are responsible for cell-mediated immunity, helper T cells are involved in activating B cells as well as other T cells. Thus, both antibody-mediated immunity and cell-mediated immunity would be affected.

3. The ointment was a good idea for the poison ivy, which caused a delayed hypersensitivity reaction, e.g., too much inflammation. For the scrape it is a bad idea, since a normal amount of inflammation is beneficial and helps to fight infection in the scrape.

4. Because both immediate and delayed hypersensitivities produce inflammation, the fact that the metal in the jewelry resulted in inflammation is not enough information to answer the question. However, the fact that it took most of the day (many hours) to develop the reaction would indicate a delayed hypersensitivity reaction caused by T cells.

5. The drug is probably suppressing cell-mediated immunity, which is responsible for rejecting grafts and is involved in tumor control. One might expect a higher incidence of cancer in such a patient, and there is evidence that this is true for some cancers. However, it is not true for many cancers, which would indicate that other components of the immune system are also involved in tumor control.

The Respiratory System

FOCUS: The respiratory system consists of the nasal cavity, pharynx, larynx, trachea, and lungs. The diaphragm and thoracic wall muscles change the volume of the thoracic cavity, producing pressure gradients responsible for the movement of air into and out of the lungs. Once in the blood, oxygen is mostly transported bound to hemoglobin, and most carbon dioxide is transported as bicarbonate ions. Respiration is controlled by centers in the brain. The most important regulators of resting respiration are blood carbon dioxide and pH levels, although low blood oxygen levels can increase respiration. Respiration during exercise is mostly determined by the cerebral motor cortex and by feedback from proprioceptors.

CONTENT LEARNING ACTIVITY

Nose and Nasal Cavity

❝Air passing through the nasal cavity is warmed, humidified, and cleaned.❞

Match these terms with the correct statement or definition:

Conchae
Epithelium
Hard palate
Nasal septum
Paranasal sinuses

_____ 1. Divides the nasal cavity into right and left parts.

_____ 2. Forms the floor of the nasal cavity.

_____ 3. Bony ridges on the lateral walls of the nasal cavity; function to increase surface area.

_____ 4. Air-filled spaces within bones that connect to the nasal cavity.

_____ 5. Produces mucus that traps debris in the air; moves mucus to the pharynx.

Pharynx and Larynx

66 *Sometimes the pharynx is called the throat and the larynx is called the voice box.* **99**

A. Match these terms with the correct statement or definition:

Epiglottis
Laryngitis
Loudness
Pharynx

Pitch
Thyroid cartilage
Vocal cords

_____ 1. Common passageway of the respiratory and digestive systems; connects to the larynx and the esophagus.

_____ 2. Largest part of the larynx, which forms the Adam's apple.

_____ 3. Ligaments that vibrate to produce sounds.

_____ 4. Determined by the force of air moving past the vocal cords.

_____ 5. Determined by the tension of the vocal cords.

_____ 6. Inflammation of the mucous membranes of the vocal cords.

_____ 7. Covers the opening of the larynx, preventing materials from entering the larynx during swallowing.

☞ The upper respiratory tract consists of the nose, nasal cavity, and pharynx.

B. Match these terms with the correct parts labeled in Figure 15-1.

Bronchus
Larynx
Lower respiratory tract
Lungs

Nasal cavity
Pharynx
Trachea
Upper respiratory tract

1. _____

2. _____

3. _____

4. _____

5. _____

6. _____

7. _____

8. _____

Oral cavity

219

Trachea, Bronchi, and Lungs

" *The trachea and bronchi serve as passageways for air between the larynx and lungs; the lungs* **"** *are the principal organs of respiration.*

A. **M**atch these terms with the
correct statement or definition:

Alveoli
Bronchioles
Lobes
Lobules

Primary bronchi
Secondary bronchi
Tertiary bronchi
Trachea

_____ 1. Extends from the larynx and divides to form the primary bronchi; supported by C-shaped cartilages.

_____ 2. Sections of lung separated by connective tissue but not visible as surface fissures.

_____ 3. Tubes that supply each lobe of the lung.

_____ 4. Tubes that supply the lobules of the lungs.

_____ 5. Contains smooth muscle that regulates air flow.

_____ 6. Place where gas exchange takes place.

B. **M**atch these terms with the correct parts labeled in Figure 15-2:

Alveoli
Bronchiole
Primary bronchus
Secondary bronchus
Tertiary bronchus
Trachea

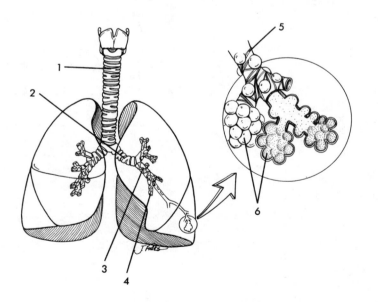

1. _____

2. _____

3. _____

4. _____

5. _____

6. _____

Pleural Cavities

"Each lung is surrounded by a separate pleural cavity.**"**

Match these terms with the
correct statement or definition:

Parietal pleura
Pleural cavity
Pleural fluid

Thoracic cavity
Visceral pleura

_____ 1. Cavity that contains the lungs and the pleural cavities.

_____ 2. Cavity formed by membranes; surround the lungs.

_____ 3. The part of the pleural membrane that is in contact with the lungs.

_____ 4. The pleural cavity contains a thin film of this substance which acts
   as a lubricant and helps to hold the pleural membranes together.

☞ Pleurisy is an inflammation of the pleural membranes.

Ventilation

"Ventilation, or breathing, is the process of moving air into and out of the lungs.**"**

Using the terms provided, complete the following statements:

Decreases
Diaphragm
Expiration
External intercostal
Higher

Increases
Inspiration
Internal intercostal
Lower

The phase of ventilation during which air is moving into the
lungs is called (1). Two principles govern the movement of
air: First, air moves from a (2) to a (3) area of pressure;
Second, as the volume of a container increases, the pressure
within the container (4). The muscle primarily responsible
for changing the volume of the thorax is the (5). When this
muscle contracts, the volume of the thorax (6), lung volume
(7), and pressure within the alveoli (8). Consequently air
pressure in the alveoli is (9) than atmospheric air and air
moves into the lungs. During quiet expiration the diaphragm
relaxes and thorax volume (10) as the thorax returns to its
resting position. Consequently air pressure in the alveoli is
(11) than atmospheric pressure and air moves out of the lungs.
In addition to the diaphragm, the intercostal muscles change
thorax volume by moving the ribs. The (12) muscles increase
thoracic volume during inspiration, and the (13) muscles
decrease thoracic volume during labored breathing.

1. _____

2. _____

3. _____

4. _____

5. _____

6. _____

7. _____

8. _____

9. _____

10. _____

11. _____

12. _____

13. _____

Understanding Lung Collapse

❝*Two factors tend to make the lungs collapse, and two factors keep the lungs from collapsing.*❞

Match these terms with the correct statement or definition:

Elastic recoil of lungs
Pleural membranes adhere to each other
Pneumothorax

Respiratory distress syndrome
Surface tension of alveolar fluid
Surfactant

_____ 1. Two factors that cause the lungs to collapse.

_____

_____ 2. Two factors that keep the lungs from collapsing.

_____

_____ 3. Mixture of lipoproteins produced by the epithelium of the alveoli; reduces surface tension.

_____ 4. Results when newborns do not manufacture enough surfactant and the lungs tend to collapse.

_____ 5. Entry of air into the pleural cavity; causes the lung to collapse.

Pulmonary Volumes

❝*A spirometer is a device for measuring the volumes of air that move into and out of the lungs.*❞

A. Match these terms with the correct statement or definition:

Expiratory reserve volume
Inspiratory reserve volume
Residual volume

Tidal volume
Vital capacity

_____ 1. Volume of air inspired or expired by quiet breathing.

_____ 2. Volume of air that can be taken in after the inspiration of a normal tidal volume.

_____ 3. Volume of air that can be blown out after the expiration of a normal tidal volume.

_____ 4. Volume of air in the respiratory passages and lungs after maximum expiration.

_____ 5. The sum of the inspiratory reserve volume, tidal volume, and expiratory reserve volume.

_____ 6. The volume of air that a person can expel from the respiratory tract after a maximum inspiration.

B. Match these terms with the correct parts labeled in Figure 15-3:

Expiratory reserve volume   Tidal volume
Inspiratory reserve volume   Vital capacity
Residual volume

1. _____

2. _____

3. _____

4. _____

5. _____

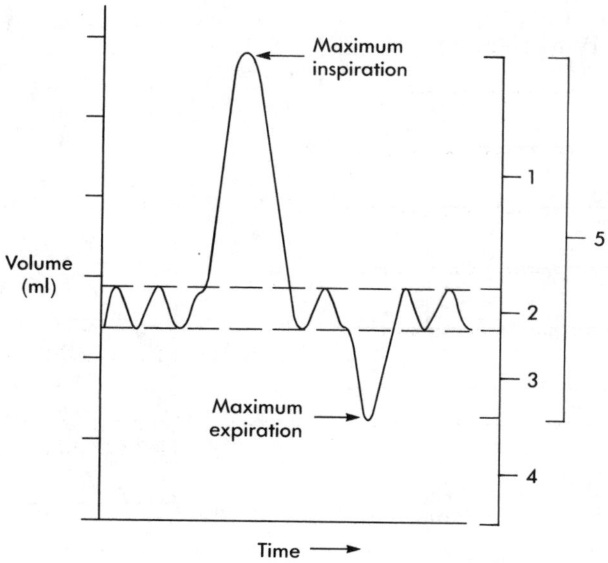

Gas Exchange

❝*Gas exchange takes place across the respiratory membrane.***❞**

A. Match these terms with the correct statement or definition:

Alveoli   Increases
Blood   Respiratory membrane
Decreases

_____ 1. Structure across which gas exchange between air and blood takes place.

_____ 2. Has the highest concentration of oxygen.

_____ 3. Effect on gas exchange when the respiratory membrane becomes thicker; an example is pneumonia.

_____ 4. Effect on gas exchange when the surface area of the respiratory membrane decreases; an example is emphysema.

B. Match these terms with
the correct parts labeled
in Figure 15–4:

Alveolar wall
Capillary wall
Fluid with surfactant
Interstitial space
Respiratory membrane

1. _____

2. _____

3. _____

4. _____

5. _____

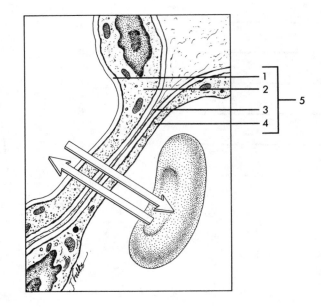

Oxygen and Carbon Dioxide Transport in the Blood

"_Oxygen is transported bound to hemoglobin; carbon dioxide is mostly transported as bicarbonate ion._**"**

Using the terms provided, complete the following statements:

Bicarbonate ions   Hemoglobin
Blood proteins   Hydrogen ions
Carbonic anhydrase   Increases
Decreases   Plasma

Approximately 97% of oxygen is transported by _(1)_. The
remaining 3% of oxygen is transported dissolved in _(2)_.
Approximately 8% of carbon dioxide is transported by _(3)_,
20% by _(4)_ (primarily hemoglobin), and 72% as _(5)_. The
enzyme _(6)_ inside red blood cells _(7)_ the rate at which carbon
dioxide and water react to form _(8)_ and bicarbonate ions.
Consequently, when carbon dioxide levels increase, hydrogen
ion levels increase, and blood pH _(9)_.

1. _____

2. _____

3. _____

4. _____

5. _____

6. _____

7. _____

8. _____

9. _____

Control of Respiration

Respiration is regulated by the respiratory center in the medulla oblongata and lower pons.

Match these terms with the correct statement or definition:

Carotid and aortic bodies
Decrease
Increase
Medulla oblongata

Motor cortex
Proprioceptors
Respiratory center

_____ 1. Part of brain that rhythmically stimulates the muscles of inspiration such as the diaphragm.

_____ 2. Chemoreceptors located here are most sensitive to small changes in blood carbon dioxide and pH levels.

_____ 3. Chemoreceptors located here primarily respond to low oxygen levels.

_____ 4. Change in respiration rate that occurs as a result of a small decrease in blood carbon dioxide.

_____ 5. Change in respiration rate that occurs as a result of a small decrease in blood pH.

_____ 6. Change in respiration rate that occurs as a result of a large decrease in blood oxygen.

_____ 7. Respiratory center is most affected by these two during exercise.

_____

Disorders of the Respiratory System

Knowledge of respiratory disorders is clinically important.

A. Match these terms with the correct statement or definition:

Asthma
Bronchitis
Emphysema
Lung cancer

Paralysis
Pulmonary fibrosis
Sudden infant death syndrome

_____ 1. Inflammation of the bronchi.

_____ 2. Results in the destruction of the walls of the alveoli.

_____ 3. Contraction of the bronchioles; often caused by allergic reactions.

_____ 4. Exposure to asbestos, silica, or coal dust result in the replacement of normal lung tissue with fibrous connective tissue.

_____ 5. Tumor arising in the epithelium of the respiratory tract.

_____ 6. Causes death when infants stop breathing during sleep.

_____ 7. Inability to contract respiratory muscles; caused by trauma to the nervous system or the polio virus.

B. Match these terms with the correct statement or definition:

Common cold   Pneumonia
Diphtheria   Strept throat
Flu   Tuberculosis
Fungal infections   Whooping cough

_____ 1. Bacterial infection characterized by inflammation of the pharynx and fever; treated effectively with antibiotics.

_____ 2. Bacterial infection producing a membrane that blocks the respiratory passages.

_____ 3. Viral infection that typically causes sneezing, excessive nasal secretions, and congestion; usually last a week.

_____ 4. Bacterial infection that causes a loss of cilia from respiratory epithelial cells; mucus builds up in the respiratory passages.

_____ 5. Bacterial infection that produces tubercles in the lungs.

_____ 6. General term for many lung infections; caused by many different kinds of microorganisms and characterized by accumulation of fluid in the alveoli.

_____ 7. Viral infection characterized by chills, fever, headaches, and muscular aches.

_____ 8. Respiratory infection contracted from spores in dust or animal feces.

QUICK RECALL

1. Trace the path of inspired air from the outside of the body through the upper respiratory tract.

2. Trace the path of inspired air through the lower respiratory tract.

3. State the two physical principles that are involved with the movement of air into and out of the lungs.

4. Name the four lung volumes.

5. List two factors that tend to cause the lungs to collapse and two factors that prevent the lungs from collapsing.

6. List the four parts of the respiratory membrane through which gases must diffuse.

7. Rank in order of importance the ways in which oxygen and carbon dioxide are transported.

8. List three chemical factors that influence respiration, the location in the body where the levels of these chemicals are monitored, and the changes of these chemicals that cause an increase in respiration rate.

MASTERY LEARNING ACTIVITY

Place the letter corresponding to the correct answer in the space provided.

_____ 1. The nasal cavity
    a. warms and humidifies air.
    b. is divided into left and right parts by the hard palate.
    c. contains the openings to the auditory tubes.
    d. all of the above

_____ 2. The pharynx
    a. is closed off by the epiglottis when materials are swallowed.
    b. opens into the oral and nasal cavities.

    c. has openings from the paranasal sinuses.
    d. all of the above

_____ 3. The larynx
    a. connects the pharynx to the esophagus.
    b. has C-shaped cartilages.
    c. contains the vocal cords.
    d. all of the above

_____ 4. Which of the following parts of the respiratory passages is correctly matched with the structure it supplies?
a. trachea - secondary bronchi
b. primary bronchi - lobes of the lungs
c. secondary bronchi - lobules of the lungs
d. bronchioles - alveoli

_____ 5. During an asthma attack, the patient has difficulty breathing because of constriction of the
a. trachea.
b. bronchi.
c. bronchioles.
d. alveoli.

_____ 6. The parietal pleura
a. covers the surface of the lungs.
b. lines the inner surface of the thoracic cavity.
c. separates lobules of the lungs from each other.
d. is the membrane across which gas exchange occurs.

_____ 7. When the diaphragm contracts, the
a. volume of the thoracic cavity increases.
b. air pressure in the lungs decreases below atmospheric pressure.
c. air moves into the lungs.
d. all of the above

_____ 8. The lungs do not normally collapse because of
a. surfactant.
b. elastic recoil of lung tissue.
c. surface tension of alveolar fluid.
d. all of the above

_____ 9. A patient expires normally; then, using forced ventilation, he blows as much air as possible into a spirometer. This would measure the
a. inspiratory reserve.
b. expiratory reserve.
c. residual volume.
d. tidal volume.
e. vital capacity.

_____ 10. Which of the following layers must gases cross to pass from the air to the blood?
a. simple squamous epithelium
b. thin interstitial space
c. thin layer of alveolar fluid
d. all of the above

_____ 11. Gases diffuse across the respiratory membrane
a. from areas of higher to areas of lower gas concentration.
b. more readily when the thickness of the respiratory membrane is increased.
c. less readily when the surface area of the respiratory membrane is increased.
d. all of the above

_____ 12. Most carbon dioxide is transported in the blood
a. dissolved in plasma.
b. bound to blood proteins, primarily hemoglobin.
c. within bicarbonate ions.
d. bound to iron.

_____ 13. Blood oxygen levels
a. are more important than carbon dioxide levels in the regulation of respiration.
b. need to change only slightly to cause a change in respiration.
c. are monitored primarily by chemoreceptors in the carotid and aortic bodies.
d. all of the above

_____ 14. During exercise, respiration rate and depth increase primarily because of
a. increased blood carbon dioxide levels.
b. decreased blood oxygen levels.
c. decreased blood pH
d. input to the respiratory center from the cerebral motor cortex and from proprioceptors.

Use a separate sheet of paper to complete this section.

1. One technique for artificial respiration is mouth-to-mouth resuscitation. The rescuer takes a deep breath, blows air into the victim's mouth, and lets air flow out of the victim. The process is repeated. Explain the following: (1) Why do the victim's lungs expand; (2) why does air move out of the victim's lungs?

2. Marty Blowhard used a spirometer with the following results:
   a. After a normal inspiration, a normal expiration was 500 ml.
   b. Following a normal expiration, he was able to expel an additional 1000 ml.
   c. Taking as deep a breath as possible, then forcefully exhaling all the air possible, yielded an output of 4500 ml.

   On the basis of these measurements, what is Marty's inspiratory reserve?

3. Predict what would happen to tidal volume if the phrenic nerves were cut; if the spinal nerves to the intercostal muscles were cut.

4. A patient has severe emphysema that has extensively damaged the alveoli and reduced the surface area of the respiratory membrane. Although the patient is receiving oxygen therapy, he still has a tremendous urge to take a breath, that is, he does not feel as if he is getting enough air. Explain why this occurs.

5. A skin diver hyperventilates, takes a deep breath, and dives under the water. Unfortunately, the diver passes out and drowns. Explain how this could happen.

ANSWERS TO CHAPTER 15

CONTENT LEARNING ACTIVITY

Nose and Nasal Cavity
1. Nasal septum; 2. Hard palate; 3. Conchae;
4. Paranasal sinuses; 5. Epithelium

Pharynx and Larynx
A. 1. Pharynx; 2. Thyroid cartilage; 3. Vocal cords;
4. Loudness; 5. Pitch; 6. Laryngitis; 7. Epiglottis
B. 1. Nasal cavity; 2. Pharynx; 3. Upper respiratory tract; 4. Larynx; 5. Trachea; 6. Bronchus; 7. Lungs;
8. Lower respiratory tract

Trachea, Bronchi, and Lungs
A. 1. Trachea; 2. Lobules; 3. Secondary bronchi;
4. Tertiary bronchi; 5. Bronchioles; 5. Alveoli
B. 1. Trachea; 2. Primary bronchus; 3. Secondary bronchus; 4. Tertiary bronchus; 5. Bronchiole;
6. Alveoli

Pleural Cavities
1. Thoracic cavity; 2. Pleural cavity; 3. Visceral pleura;
4. Pleural fluid

Ventilation
1. Inspiration; 2. Higher; 3. Lower; 4. Decreases;
5. Diaphragm; 6. Increases; 7. Increases;
8. Decreases; 9. Lower; 10. Decreases; 11. Higher;
12. External intercostal; 13. Internal intercostal

Understanding Lung Collapse
1. Elastic recoil of lungs and surface tension of alveolar fluid; 2. Pleural membranes adhere to each other and surfactant; 3. Surfactant; 4. Respiratory distress syndrome; 5. Pneumothorax

Pulmonary Volumes
A. 1. Tidal volume; 2. Inspiratory reserve volume;
3. Expiratory reserve volume; 4. Residual volume;
5. Vital capacity; 6. Vital capacity
B. 1. Inspiratory reserve volume; 2. Tidal volume;
3. Expiratory reserve volume; 4. Residual volume;
5. Vital capacity

Gas Exchange
- A. 1. Respiratory membrane; 2. Alveoli; 3. Decreases; 4. Decreases
- B. 1. Fluid with surfactant; 2. Alveolar wall; 3. Interstitial space; 4. Capillary wall; 5. Respiratory membrane

Oxygen and Carbon Dioxide Transport in the Blood
1. Hemoglobin; 2. Plasma; 3. Plasma; 4. Blood proteins; 5. Bicarbonate ions; 6. Carbonic anhydrase; 7. Increases; 8. Hydrogen ions; 9. Decreases

Control of Respiration
1. Respiratory center; 2. Medulla oblongata; 3. Carotid and aortic bodies; 4. Decrease; 5. Increase; 6. Increase; 7. Motor cortex and proprioceptors

Disorders of the Respiratory System
- A. 1. Bronchitis; 2. Emphysema; 3. Asthma; 4. Pulmonary fibrosis; 5. Lung cancer; 6. Sudden infant death syndrome; 7. Paralysis
- B. Strept throat; 2. Diphtheria; 3. Common cold; 4. Whooping cough; 5. Tuberculosis; 6. Pneumonia; 7. Flu; 8. Fungal infections

QUICK RECALL

1. Nostrils, nasal cavity, and pharynx
2. Larynx, trachea, primary bronchus, secondary bronchus, tertiary bronchus, bronchiole, alveolus
3. Air moves through a tube from an area of higher to an area of lower pressure; as volume increases pressure decreases, and as volume decreases, pressure increases
4. Pulmonary volumes: tidal volume, inspiratory reserve volume, expiratory reserve volume, residual volume
5. The lungs tend to collapse because of the elastic recoil of the lungs and surface tension of alveolar fluid. The lungs are prevented from collapsing by adhesion of the pleural membranes to each other and by surfactant.

6. Fluid lining the alveolar wall, the alveolar wall, an interstitial space, and the capillary wall
7. Oxygen: 97% bound to hemoglobin, 3% dissolved in plasma; Carbon dioxide: 72% as bicarbonate ions, 20% bound to blood proteins (primarily hemoglobin), 8% dissolved in plasma
8. Carbon dioxide, pH (hydrogen ions), and oxygen. Chemoreceptors in the medulla oblongata are most sensitive to small changes in carbon dioxide and pH. An increase in carbon dioxide or a decrease in pH stimulates respiration. Chemoreceptors in the carotid and aortic bodies are most sensitive to changes in oxygen. A large decrease in oxygen stimulates respiration.

MASTERY LEARNING ACTIVITY

1. A. The nasal cavity warms and humidifies air. It also filters the air removing debris. The nasal septum divides the nasal cavity into left and right parts; the hard palate forms the floor of the nasal cavity and separates the nasal cavity from the oral cavity. The auditory tubes open into the pharynx.

2. B. The pharynx is the common passageway for the respiratory system and the digestive system. The larynx is closed off by the epiglottis during swallowing. The paranasal sinuses open into the nasal cavity.

3. C. The larynx contains the vocal cords. C-shaped cartilages are part of the trachea. The larynx connects the pharynx to the trachea.

4. D. The bronchioles supply the alveoli. The trachea divides to from the primary bronchi, which supply the lungs. The secondary bronchi supply the lobes of the lungs, and the tertiary bronchi supply the lobules of the lungs.

5. C. The walls of the bronchioles are smooth muscle that can constrict and impede air flow during an asthma attack. The trachea and bronchi are held open by C-shaped cartilages.

6. B. The parietal pleura lines the inner surface of the thoracic cavity and the visceral pleura covers the surfaces of the lungs. Gas exchange takes place across the respiratory membrane.

7. D. Contraction of the diaphragm increases thoracic volume, which causes a decrease in air pressure within the lungs. Air moves from outside the body (higher pressure) to the lower pressure inside the lungs.

8. A. Surfactant lowers the surface tension of the alveolar fluid and helps to prevent collapse of the lungs. Elastic recoil of lung tissue and surface tension of alveolar fluid are responsible for collapse of the lungs.

9. B. By definition this is expiratory reserve.

10. D. Gases must pass through the parts of the respiratory membrane: thin layer of alveolar fluid, alveolar wall (simple squamous epithelium), thin interstitial space, and capillary wall (simple squamous epithelium).

11. A. Gases diffuse from areas of higher to areas of lower gas concentration. Diffusion decreases when the thickness of the respiratory membrane is increased, and increases when the surface area of the respiratory membrane is increased.

12. C. Most (72%) carbon dioxide is transported as bicarbonate ions. It is also transported bound to blood proteins (20%) such as hemoglobin, and dissolved in plasma (8%). Iron, within the heme portion of hemoglobin, transports oxygen.

13. C. The chemoreceptors most sensitive to oxygen are in the carotid and aortic bodies. However, it takes a large change in oxygen to produce a change in respiration through these receptors. The chemoreceptors in the medulla oblongata are most sensitive to small changes in carbon dioxide or pH, which are the chemical factors most important in the regulation of respiration.

14. D. Input from the cerebral motor cortex and from proprioceptors stimulates the respiratory center and causes an increase in respiration rate and depth during exercise. Blood oxygen and carbon dioxide levels do not change much during exercise if respiration is matched to the amount of exercise being performed.

 ★ FINAL CHALLENGES ★

1. The victim's lungs expand because of the pressure generated by the rescuer. This fills the victim's lungs with air that has a greater pressure than atmospheric pressure. Air flows out of the victim's lungs because of this pressure difference and because of the recoil of the victim's lungs.

2. The inspiratory reserve volume is 3000 ml. It is equal to the vital capacity minus the sum of the tidal volume and the expiratory reserve, that is, 4500 - (500 + 1000) = 3000.

3. The phrenic nerves supply the diaphragm. Cutting these nerves paralyzes the diaphragm and tidal volume decreases so greatly that death probably results. Cutting the spinal nerves to the intercostal muscles inhibits movement of the ribs. This decreases tidal volume unless the diaphragm compensates.

4. The reduced surface area of the respiratory reduces oxygen and carbon dioxide gas exchange. Administering oxygen increases the concentration of oxygen in the alveoli and promotes the exchange of oxygen. Carbon dioxide exchange, however, is not increased by the oxygen therapy. Instead, blood carbon dioxide levels increase and stimulate the respiratory center. The patient feels as if he is not getting enough air.

5. Hyperventilation decreases the amount of carbon dioxide in the blood. This makes it possible to hold one's breath for a longer period of time because it decreases the urge to take a breath. In the meantime, however, oxygen is used up. It is possible for oxygen levels to decrease enough that the diver passes out.

The Digestive System

FOCUS: The digestive system functions to ingest, digest, and absorb food and liquids. These processes provide the body with water, electrolytes, and nutrients. The digestive system consists of the digestive tract, a hollow tube extending from the mouth to the anus, plus accessory organs. Food enters the mouth and passes to the esophagus, stomach, small intestine, large intestine, and rectum, with undigested food exiting through the anus. Regulation of digestive tract functions is accomplished by the nervous system and hormones.

CONTENT LEARNING ACTIVITY

Anatomy and Histology of the Digestive System
"_Nearly all portions of the digestive tube consist of four layers._**"**

A. Match these tunics with the correct description or definition:

Mucosa   Serosa
Muscularis   Submucosa
Peristalsis

_____ 1. Innermost layer; mucous membrane thickened in places to resist abrasion, thin in areas of absorption and secretion.

_____ 2. Layer just outside the mucosa; a thick layer of loose connective tissue containing nerves, blood vessels, and small glands.

_____ 3. Layer composed of circular and longitudinal layers of smooth muscle.

_____ 4. Outermost layer; composed of epithelium or connective tissue.

_____ 5. Wavelike movements that push materials along the digestive tract.

 Neurons within the submucosa and muscularis form the intramural plexus, which is important for control of movement and secretion in the digestive tract.

B. Match these terms with
the correct parts labeled
in Figure 16-1:

Circular muscle
Longitudinal muscle
Mucosa
Muscularis
Serosa
Submucosa

1. _____

2. _____

3. _____

4. _____

5. _____

6. _____

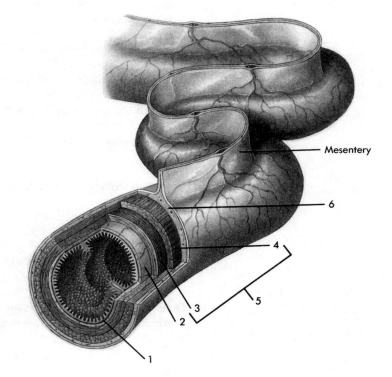

Mesentery

6

4

3

5

2

1

Oral Cavity

❝The oral cavity, or mouth, is the first portion of the digestive tract.❞

A. Using the terms provided, complete the following statements:

Speech   Teeth
Taste   Tongue

The oral cavity, or mouth, is bounded by the lips and cheeks,
and contains the (1) and tongue. The lips, cheeks, and tongue
are important in the processes of food chewing and (2). The
(3) is a large, muscular organ that occupies most of the oral
cavity. The tongue is important in swallowing and is a major
sensory organ for (4).

1. _____

2. _____

3. _____

4. _____

B. Match these terms with the correct statement or definition:

Crown   Neck
Dentin   Periodontal ligaments
Enamel   Pulp
Gingiva (gums) Pulp cavity

_____ 1. Cutting or chewing surface with one or more cusps (points).

_____ 2. Part of the tooth between the crown and the root.

_____ 3. Space in the center of the tooth.

_____ 4. Blood vessels, nerves, and connective tissue located in the pulp cavity.

_____ 5. Living, cellular, calcified tissue surrounding the pulp cavity.

_____ 6. Extremely hard, acellular substance that protects the tooth against acids and abrasion.

_____ 7. Connective tissue that holds the roots of teeth in place in sockets in the mandible and maxillae.

_____ 8. Dense fibrous connective tissue and moist stratified squamous epithelium that covers the mandible and maxilla.

C. Match these numbers with the correct statement:

One   Two
Three

_____ 1. Number of incisors in each of four parts of the mouth.

_____ 2. Number of canines in each of four parts of the mouth.

_____ 3. Number of premolars in each of four parts of the adult mouth.

_____ 4. Number of molars in each of four parts of the adult mouth.

D. Match these terms with the correct statement or definition:

Incisors and canines   Primary teeth
Molars and premolars   Secondary teeth

_____ 1. Teeth of the adult mouth.

_____ 2. Deciduous teeth; also called milk teeth.

_____ 3. Primarily cutting and tearing teeth.

_____ 4. Primarily grinding teeth.

 Food taken into the mouth is chewed, or masticated by the teeth.

E. **Match these terms with the correct parts labeled in Figure 16-2:**

Crown   Neck
Cusp   Periodontal ligaments
Dentin   Pulp cavity
Enamel   Root

1. _____

2. _____

3. _____

4. _____

5. _____

6. _____

7. _____

8. _____

F. **Match these terms with the correct statement or definition:**

Hard palate   Uvula
Soft palate

_____ 1. Anterior bony portion of the roof of the oral cavity.

_____ 2. Posterior portion of the roof of the oral cavity, composed of skeletal muscle and connective tissue.

_____ 3. Projection hanging from the posterior edge of the soft palate.

G. **Match these terms with the correct statement or definition:**

Lysozyme   Salivary amylase
Parotid glands   Sublingual glands
Saliva   Submandibular glands

_____ 1. The largest salivary glands; located just anterior to each ear.

_____ 2. Salivary glands located along the inferior border of the mandible.

_____ 3. Mixture of mucus and serous (watery) fluids that contains digestive enzymes.

_____ 4. A digestive enzyme found in the serous portion of saliva.

_____ 5. An enzyme that causes some bacterial cells to rupture.

Pharynx and Esophagus

"_The esophagus is a muscular tube that extends from the pharynx (throat) to the stomach._**"**

Using the terms provided, complete the following statements:

Bolus   Peristaltic
Epiglottis   Pharynx
Esophageal sphincter   Reflex
Heartburn   Soft palate
Muscles

The esophagus transports food from the _(1)_ to the stomach.
Swallowing is partly a voluntary activity, and partly an
involuntary _(2)_ . During the voluntary phase, a _(3)_ , or mass
of food is formed in the mouth and pushed into the pharynx.
Pharyngeal _(4)_ contract, and food is pushed into the
esophagus. As food passes through the pharynx, the
_(5)_ closes off the openings into the nasal cavity and the
_(6)_ covers the opening into the larynx. Muscular contractions
of the esophagus occur in _(7)_ waves, and the bolus is propelled
through the esophagus. When a bolus reaches the end of the
esophagus, the _(8)_ relaxes and the bolus enters the stomach.
If stomach acid refluxes into the esophagus, it can produce the
unpleasant symptoms of _(9)_ .

1. _____

2. _____

3. _____

4. _____

5. _____

6. _____

7. _____

8. _____

9. _____

Stomach

"_The stomach is an enlarged segment of the digestive tract with three parts._**"**

A. Match these terms with the
correct statement or definition:

Body   Pylorus
Fundus

_____ 1. Portion of the stomach to the left and superior to the opening of the
esophagus.

_____ 2. The central portion of the stomach.

_____ 3. The inferior part of the stomach that connects to the small intestine.

☞ The pyloric sphincter is a thick ring of smooth muscle that surrounds the opening from the
stomach into the small intestine.

B. Match these terms with the correct parts labeled in Figure 16-3:

Body
Esophageal sphincter
Fundus
Pylorus
Pyloric sphincter
Rugae

1. _____

2. _____

3. _____

4. _____

5. _____

6. _____

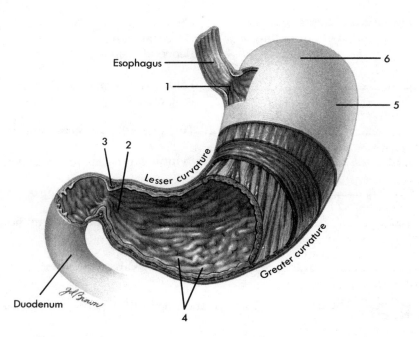

C. Using the terms provided, complete the following statements:

Chyme   Pepsin
Hydrochloric acid Rugae
Intrinsic factor   Smooth muscle
Gastric juice   Storage
Mucus

The muscular layer of the stomach is different from other regions of the GI tract in that it consists of three layers of (1). When the stomach is empty, the submucosa and mucosa of the stomach are thrown into large folds called (2), which allow the stomach to stretch. On the stomach's inner surface, the epithelium produces (3) that coats and protects the stomach lining from stomach acid and digestive enzymes. The epithelium is folded to form many tube-shaped gastric glands that produce (4). Gastric juice contains mucus, (5) (an enzyme that begins protein digestion), (6) (provides the proper pH environment for pepsin), and (7) (increases absorption of vitamin B_{12}). The stomach functions primarily as a (8) and mixing chamber for food. As food enters the stomach, food and gastric juice are mixed to become a semifluid mixture called (9).

1. _____

2. _____

3. _____

4. _____

5. _____

6. _____

7. _____

8. _____

9. _____

Small Intestine

The small intestine is about 6 meters long and consists of the duodenum, jejunum, and ileum.

A. **M**atch these terms with the correct statement or definition:

Circular folds	Lacteals	
Ileocecal valve	Microvilli	
Ileocecal sphincter	Villi	

_____ 1. Folds in mucosal and submucosal layers that run perpendicular to the long axis of the digestive tract.

_____ 2. Tiny fingerlike projections of the mucosa.

_____ 3. Lymph capillaries found in villi.

_____ 4. Tubelike extensions of the cell membrane on the surface of villi.

_____ 5. Ring of smooth muscle surrounding the junction between the small and large intestine.

_____ 6. One-way valve at the junction between the ileum and small intestine.

☞ The circular folds, villi, and microvilli function to increase surface area in the small intestine.

B. **U**sing the terms provided, complete the following statements:

Absorption   Microvilli
Chyme   Mucus
Duodenum   Pancreas
Enzymes   Smooth muscle

The small intestine, especially the _(1)_ and the jejunum, are the primary sites of digestion and _(2)_ in the GI tract. Secretions of the _(3)_ and liver enter the duodenum and continue the process of digestion begun by saliva and gastric juices. The epithelium lining the small intestine also produces _(4)_ that complete the digestive process, and the digested food molecules are absorbed through the _(5)_. The mucosa of the small intestine and intestinal glands also produce secretions that primarily contain _(6)_, electrolytes, and water. Intestinal secretions lubricate and protect the intestinal wall from acidic _(7)_ and action of digestive enzymes. Contraction of _(8)_ within the wall of the small intestine mix secretions with chyme and move the mixture to the large intestine.

1. _____

2. _____

3. _____

4. _____

5. _____

6. _____

7. _____

8. _____

Large Intestine

66 *The large intestine consists of the cecum, colon, rectum, and anal canal.* **99**

A. Match these terms with the correct statement or definition:

Anal canal	Colon
Anus	External anal sphincter
Appendix	Internal anal sphincter
Cecum	Rectum

_____ 1. Blind sac that extends inferiorly past the junction of the small and large intestine.

_____ 2. Small blind tube attached to the cecum.

_____ 3. Part of the large intestine that consists of ascending, transverse, descending, and sigmoid portions.

_____ 4. Straight muscular tube between the sigmoid colon and the anal canal.

_____ 5. The last 2 to 3 cm of the digestive tract.

_____ 6. External GI tract opening.

_____ 7. Smooth muscle layer; encircles the anal canal at the superior end.

_____ 8. Skeletal muscle layer; encircles the inferior end of the anal canal.

B. Using the terms provided, complete the following statements:

Defecation	Reproduce
Feces	Vitamin K
Microorganisms	Water
Mucus	

While in the colon, chyme is converted to _(1)_. Absorption of _(2)_ and salts, secretion of _(3)_, and the extensive action of _(4)_ are involved in the formation of feces. Microorganisms in the colon _(5)_ rapidly, and ultimately compose approximately 30% of the dry weight of feces. Some bacteria synthesize _(6)_, which is passively absorbed in the colon. Feces are stored in the colon until they are eliminated by the process of _(7)_.

1. _____

2. _____

3. _____

4. _____

5. _____

6. _____

7. _____

Liver and Gallbladder

"The liver is an amazing organ that performs important digestive and excretory functions, stores and processes nutrients, synthesizes new molecules, and detoxifies harmful chemicals.

A. Match these terms with the correct statement or definition:

Common bile duct   Hepatic artery
Common hepatic duct   Hepatic portal vein
Cystic duct   Hepatic veins

_____ 1. Vessel that brings oxygen-rich blood to the liver.

_____ 2. Vessel that carries blood from the digestive tract to the liver.

_____ 3. Vessels through which blood exits the liver and enters the general circulation.

_____ 4. Passageway for transport of bile from the liver.

_____ 5. Bile passageway from the gallbladder.

_____ 6. Passageway formed from joining of common hepatic duct and cystic duct; enters the duodenum.

B. Using the terms provided, complete the following statements:

Absorbed   Bile salts
Bicarbonate ions   Bilirubin
Bile   Detoxify

1. _____

2. _____

3. _____

4. _____

5. _____

6. _____

Blood from the digestive tract is rich in (1) materials. Liver cells process nutrients and (2) harmful substances in the blood. One major function of the liver is (3) production. Although bile contains no digestive enzymes, the (4) in bile play a role in digestion by diluting and neutralizing stomach acid. In addition, (5) in bile increase the efficiency of fat digestion and absorption. Bile also contains excretory products such as (6) to be eliminated from the body.

Pancreas

66 *The pancreas is a complex organ composed of both endocrine and exocrine tissues.* **99**

Using the terms provided, complete the following statements:

Bicarbonate ion   Glucose
Endocrine   Insulin
Enzymes   Pancreatic duct
Exocrine

The _(1)_ portion of the pancreas consists of pancreatic islets. The islet cells produce _(2)_ and glucagon, hormones that are important in controlling blood levels of nutrients such as _(3)_ and amino acids. The _(4)_ portion of the pancreas consists of glands that produce digestive _(5)_. Pancreatic secretions also have a high _(6)_ content that neutralizes the acidic chyme entering the duodenum. The exocrine glands are connected by a series of ducts that join to form the _(7)_, which joins the common bile duct and empties into the duodenum.

1. _____

2. _____

3. _____

4. _____

5. _____

6. _____

7. _____

Peritoneum

66 *The body walls and organs of the abdominal cavity are lined with serous membranes.* **99**

Match these terms with the correct statement or definition:

Greater omentum   Parietal peritoneum
Lesser omentum   Retroperitoneal organs
Mesenteries   Visceral peritoneum

_____   1. Serous membrane that covers the internal organs.

_____   2. Serous membrane that covers the interior of the body wall.

_____   3. Connective tissue sheets composed of two layers of serous membrane and connective tissue; hold many abdominal organs in place.

_____   4. Mesentery connecting the stomach to the liver.

_____   5. Mesentery connecting the stomach to the transverse colon.

_____   6. Abdominal organs that lie against the abdominal wall, with no mesenteries.

Regulation of the Digestive System

The autonomic system, local reflexes, and hormones regulate the activities of the digestive system.

Using the terms provided, complete the following statements:

Endocrine cells
Gastrin
Intramural plexus
Local reflexes

Medulla oblongata
Parasympathetic
CNS Reflexes
Sympathetic

The autonomic nervous system has regulatory centers in the (1) that control the digestive system. The (2) division of the autonomic nervous system is usually stimulatory, promoting the secretions and movements necessary for digestion, whereas the (3) division usually has an inhibitory effect on the digestive system. Autonomic nervous system control is mediated through (4). For example, entry of food into the stomach stimulates stretch or chemical receptors, which causes increased gastric juice secretion and increased contraction by the stomach. Regulation also occurs by (5), which do not involve the central nervous system and are mediated by neurons in the (6). Within the epithelium of the stomach and small intestine are (7), which produce hormones that can produce a response in glands or smooth muscle of the digestive tract or accessory organs. For example, entry of food into the stomach stimulates the release of (8), which increases secretions and contraction of stomach smooth muscle.

1. _____
2. _____
3. _____
4. _____
5. _____
6. _____
7. _____
8. _____

Understanding Defecation

Elimination of feces from the body is called defecation, or a bowel movement.

Using the terms provided, complete the following statements:

Abdominal muscles
Defecation reflex
External anal sphincter
Internal anal sphincter

Intramural plexus
Mass movements
Rectum

Defecation begins with movement of feces into the (1); this movement has a voluntary and involuntary component. Voluntary actions include a large inspiration of air, closure of the larynx, and forceful contraction of the (2). The involuntary movement of feces into the rectum results from (3), which are strong peristaltic contractions of the transverse and descending colon. The contractions of mass movements are integrated in the (4) of the digestive tract; stretch of the stomach results in action potentials that travel to the colon, where smooth muscles are stimulated to contract. Distention of the rectal wall by feces initiates the (5). As feces is pushed toward the anus, the internal and external anal sphincters relax. Voluntary control of the (6), however, can block the defecation reflex and prevent defecation.

1. _____
2. _____
3. _____
4. _____
5. _____
6. _____

Digestion and Absorption

" *Digestion and absorption provide the body with needed nutrients, water, and electrolytes.* **"**

A. Match these terms with the correct statement or definition:

Absorption
Digestion

_____

_____

1. Chemical breakdown of organic molecules into smaller parts.

2. Movement of materials from the digestive tract into the circulatory or lymphatic system.

B. Match these terms with the correct statement or definition:

Amylases   Monosaccharides
Disaccharides   Polysaccharides

_____

_____

_____

_____

_____

1. Many sugar molecules joined together; e. g., starches and glycogen.

2. Two sugar molecules joined together; e. g., sucrose and lactose.

3. One sugar molecule; e. g., glucose and fructose.

4. Enzymes from saliva and pancreas that break down polysaccharides.

5. Carbohydrate molecules absorbed and transported by the blood.

C. Match these terms with the correct statement or definition:

Bile salts   Lipase
Chyle   Micelles
Emulsification   Triglycerides
Lacteals

_____

_____

_____

_____

_____

_____

_____

1. Three fatty acids bound to glycerol; the most common type of lipid.

2. Transformation of large lipid droplets into much smaller droplets.

3. Substance that emulsifies fats; secreted by the liver.

4. Enzyme secreted by the pancreas and intestine that digests lipid molecules.

5. Bile salts aggregated around small droplets of digested lipids.

6. Lymphatic capillaries in the center of villi; site of absorption of lipids.

7. Lipid-rich lymph.

D. Match these terms with the correct statement or definition:

High-density lipoproteins (HDLs)
Low-density lipoproteins (LDLs)

_____ 1. Lipoproteins that transport cholesterol to the tissues for use by cells.

_____ 2. Lipoproteins that transport cholesterol from the tissues to the liver.

_____ 3. When this lipoprotein is in excess, cholesterol is deposited in the arterial walls.

_____ 4. Aerobic exercise elevates the level of this lipoprotein in the bloodstream.

E. Match these terms with the correct statement or definition:

Amino acids   Trypsin
Pepsin

_____ 1. Stomach enzyme that breaks down proteins.

_____ 2. Protein-digesting enzyme produced by the pancreas.

_____ 3. Molecules that result from the breakdown of proteins.

F. Using the terms provided, complete the following statements:

Active transport   Hepatic portal system
Energy   Modified
Fat   Proteins

Absorption of individual amino acids occurs through the intestinal epithelial cells by _(1)_ . The _(2)_ then transports the amino acids to the liver. The amino acids may be _(3)_ in the liver, or may be released and distributed throughout the body. Amino acids are used as building blocks to form new _(4)_ , but some amino acids may be used for _(5)_ . The body cannot store amino acids, so they are partially broken down and used to synthesize glycogen or _(6)_ , which can be stored.

1. _____

2. _____

3. _____

4. _____

5. _____

6. _____

Disorders of the Digestive Tract

"*The digestive tract can be affected by many disorders.***"**

A. Match these terms with the correct statement or definition:

Dental caries   Peptic ulcer
Hiatal hernia   Periodontal disease
Malabsorption   Vomiting
Maldigestion

_____ 1. Tooth decay; results from bacteria on the tooth surface.

_____ 2. Inflammation and degeneration of the periodontal ligaments, gingiva, and bone.

_____ 3. Protrusion of the stomach through the diaphragm; may cause heartburn, difficulty in swallowing, or ulcer formation.

_____ 4. Reflex resulting from irritation, overdistention, or overexcitation of the stomach.

_____ 5. Condition in which the mucosal lining of the digestive tract is damaged and inflamed.

_____ 6. Failure of the chemical process of digestion; inadequate amounts of nutrients are available for absorption.

_____ 7. Disorders that result in abnormal nutrient absorption.

B. Match these terms with the correct statement or definition:

Appendicitis   Diarrhea
Cancer   Dysentery
Colonoscopy   Hemorrhoids
Constipation

_____ 1. Condition that occurs most commonly in the rectum and colon; may develop from polyps.

_____ 2. Procedure in which a tube is used for examination of the colon and removal of polyps.

_____ 3. Inflammation of the appendix, usually because of obstruction.

_____ 4. Slow movement of feces through the large intestine.

_____ 5. Abnormally frequent, watery bowel movements.

_____ 6. Severe form of diarrhea in which blood or mucus is present in the feces.

_____ 7. Enlarged or inflamed veins located in the wall of the anal canal.

C. Match these terms with the correct statement or definition:

Cirrhosis   Hepatitis A
Gallstones   Hepatitis B
Hepatitis   Hepatitis C

_____ 1. General name for inflammation of the liver that can occur from viral infection or alcohol consumption.

_____ 2. Mass of scar tissue in the liver.

_____ 3. Viral hepatitis usually transmitted by poor sanitation practices.

_____ 4. Two types of viral hepatitis usually transmitted by blood or body fluids.

_____

_____ 5. Viral hepatitis responsible for most cases of hepatitis following transfusions.

_____ 6. Result from precipitation of cholesterol in the gallbladder.

D. Match these terms with the correct statement or definition:

Cholera   Staphylococcal food poisoning
Giardiasis   Typhoid
Salmonellosis

_____ 1. Occurs when toxin from bacteria is ingested; appearance of symptoms is rapid.

_____ 2. Disease caused by bacteria ingested on contaminated meat, poultry, or milk; appearance of symptoms is delayed up to 36 hours.

_____ 3. A disease caused by a particularly virulent strain of *Salmonella*; spread by poor sanitation practices.

_____ 4. Disease caused by bacteria-produced toxin; causes excessive loss of fluid and electrolytes from the intestinal tract.

_____ 5. Disease caused by a protozoan that invades the intestine; commonly acquired from drinking unfiltered water from wilderness streams.

E. Match these terms with the correct statement or definition:

Ascariasis   Pinworms
Hookworms   Tapeworms

_____ 1. Intestinal parasites acquired from undercooked beef, pork, or fish; may reach lengths of 6 meters in the intestine.

_____ 2. Tiny worms that live in the digestive tract, but migrate out of the anus to lay their eggs.

_____ 3. Attach to the intestinal wall and feed on the blood and tissue of the host; spread through fecal contamination of the soil.

_____ 4. Caused by a roundworm that hatches in the upper intestine but pass into the lungs; may produce pulmonary symptoms.

QUICK RECALL

1. Name the four layers of the digestive tract.

2. List the three large pairs of salivary glands, and name the digestive enzyme in saliva.

3. List the three parts of the stomach.

4. List the names and functions of four secretions found in gastric juice.

5. List the three structural modifications that increase surface area in the small intestine.

6. List five sphincters that control movement of material through the digestive tract.

7. List four major functions of the liver in addition to the production of bile.

8. List three control mechanisms that regulate the activity of the digestive system.

Match these terms with
the correct parts labeled
in Figure 16-4:

Anal canal
Appendix
Ascending colon
Cecum
Descending colon
Esophagus
Liver
Pharynx
Rectum
Salivary glands
Sigmoid colon
Small intestine
Stomach
Transverse colon

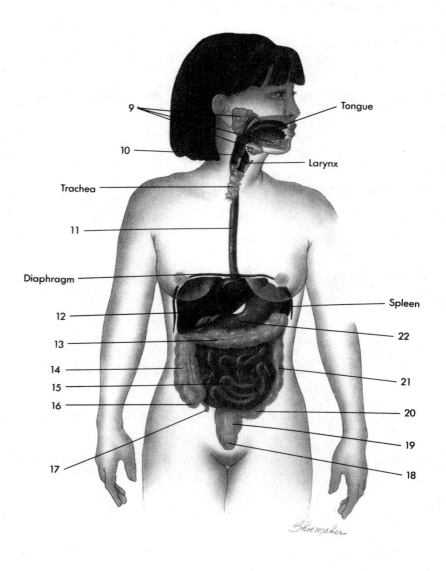

9. _____   14. _____   19. _____

10. _____   15. _____   20. _____

11. _____   16 _____   21. _____

12. _____   17. _____   22. _____

13. _____   18. _____

Place the letter corresponding to the correct answer in the space provided.

_____ 1. Which layer of the digestive tract is in direct contact with food that is consumed?
   a. mucosa
   b. muscularis
   c. submucosa
   d. serosa

_____ 2. Which of the following GI tract layers is paired with the correct description for that layer?
   a. submucosa —simple or stratified epithelium
   b. mucosa—outermost layer
   c. muscularis—circular and longitudinal layers
   d. serous—thick layer of loose connective tissue containing nerves and blood vessels
   e. all of the above

_____ 3. The tongue
   a. holds food in place during mastication.
   b. is involved in speech.
   c. assists in swallowing.
   d. is a major sensory organ for taste.
   e. all of the above

_____ 4. The number of premolar permanent teeth in each of the four parts of the mouth is
   a. 1.
   b. 2.
   c. 3.
   d. 4.

_____ 5. Dentin
   a. forms the surface of the crown of teeth.
   b. attaches the teeth to the mandible and maxilla.
   c. is found in the pulp cavity.
   d. is living, calcified, cellular material.
   e. is harder than enamel.

_____ 6. Which of the following glands secrete saliva into the oral cavity?
   a. parotid glands
   b. submandibular glands
   c. sublingual glands
   d. all of the above

_____ 7. Which of the following processes is involved in swallowing?
   a. A bolus of food is pushed into the pharynx.
   b. Constriction of pharyngeal muscles occurs.
   c. The soft palate closes off the openings into the nasal cavity.
   d. The epiglottis covers the opening into the larynx.
   e. all of the above

_____ 8. Which of the following secretions is NOT found in gastric juice?
   a. mucus
   b. hydrochloric acid
   c. intrinsic factor
   d. pepsin
   e. bicarbonate ions

_____ 9. The stomach
   a. connects to the small intestine at the fundus.
   b. has three layers of smooth muscle in the muscularis layer.
   c. when empty has mucosal folds called circular folds.
   d. functions primarily as a digestive and absorptive area of the GI tract.
   e. produces a semifluid mixture called feces.

_____10. Why doesn't the stomach digest itself?
   a. The stomach wall isn't composed of protein, so there are no digestive enzymes to attack it.
   b. The digestive enzymes in the stomach aren't strong enough.
   c. The lining of the stomach is too tough to be attacked by digestive enzymes.
   d. The stomach wall is protected by large amounts of mucus.

_____11. The function of smooth muscle contractions in the stomach is to
   a. move chyme into the small intestine.
   b. mix food with gastric juice.
   c. push food into gastric glands.
   d. both a and b

_____12. Which of the following structures function to increase the mucosal surface of the small intestine?
   a. circular folds
   b. villi
   c. microvilli
   d. all of the above

_____13. Given the following parts of the small intestine:
   1. duodenum
   2. ileum
   3. jejunum

   Choose the arrangement that lists the parts in the order food would encounter them as the food passes through the small intestine.
   a. 1,2,3
   b. 1,3,2
   c. 2,1,3
   d. 2,3,1

_____14. The small intestine
   a. is the primary site for digestion and absorption in the GI tract.
   b. has a pyloric sphincter at the junction with the large intestine.
   c. uses pancreatic enzymes for digestion, but produces no enzymes.
   d. has three parts; the duodenum is the longest.

_____15. The liver
   a. receives blood from the hepatic artery.
   b. receives blood from the hepatic portal vein.
   c. exits blood through the hepatic veins.
   d. all of the above

_____16. Which of the following would occur if a person suffered from a severe case of hepatitis that impaired liver function?
   a. Fat digestion may be hampered.
   b. Bilirubin may accumulate in the blood.
   c. harmful substances may accumulate in the blood.
   d. b and c
   e. all of the above

_____17. The exocrine gland secretions of the pancreas
   a. are produced in the pancreatic islets.
   b. contain bicarbonate ions.
   c. contain insulin and glucagon.
   d. all of the above

_____18. The portion of the digestive tract in which digestion begins is the
   a. oral cavity.
   b. esophagus.
   c. stomach.
   d. duodenum.
   e. jejunum.

_____19. Which of the following secretions is NOT correctly matched with the location where it is produced?
   a. amylase - salivary glands
   b. pepsin - small intestine
   c. trypsin - pancreas
   d. bicarbonate ions - liver and pancreas
   e. mucus - salivary glands, stomach, small intestine, large intestine

_____20. Defecation
   a. can be initiated by stretch of the rectum.
   b. can occur as a result of mass movements.
   c. involves local reflexes.
   d. involves parasympathetic reflexes
   e. all of the above

Use a separate sheet of paper to complete this section.

1. When you chew bread for a few minutes, it tastes sweet. Explain how this happens.

2. Many people have a bowel movement following a meal, especially after breakfast. Why does this occur?

3. A woman had her gallbladder removed, and noticed that every time she had a hamburger and fries at her favorite "fast food" restaurant, she had severe gastric distress. What would be your explanation to her?

4. A friend advises you that he swallows several kinds of pills each morning. He further advises you that these pills contain enzymes that help him to digest his food, and that they are worth the considerable amount of money he spends on them. What would your reply to him be? (Hint: Remember all enzymes are proteins.)

5. Why might cutting the parasympathetic nerve supply (vagus nerve) to the stomach help someone with a peptic ulcer? What side effects might be expected from such a procedure?

ANSWERS TO CHAPTER 16

CONTENT LEARNING ACTIVITY

Anatomy and Histology of the Digestive System
A. 1. Mucosa; 2. Submucosa; 3. Muscularis;
   4. Serosa; 5. Peristalsis
B. 1. Mucosa; 2. Submucosa; 3. Circular muscle;
   4. Longitudinal muscle; 5. Muscularis; 6. Serosa

Oral Cavity
A. 1. Teeth; 2. Speech; 3. Tongue; 4. Taste
B. 1. Crown; 2. Neck; 3. Pulp cavity; 4. Pulp;
   5. Dentin; 6. Enamel; 7. Periodontal ligaments;
   8. Gingiva (gums)
C. 1. Two; 2. One; 3. Two; 4. Three
D. 1. Secondary teeth; 2. Primary teeth; 3. Incisors
   and canines; 4. Molars and premolars
E. 1. Cusp; 2. Enamel; 3. Dentin; 4. Pulp cavity;
   5. Periodontal ligaments; 6. Root; 7. Neck;
   8. Crown
F. 1. Hard palate; 2. Soft palate; 3. Uvula
G. 1. Parotid glands; 2. Submandibular glands;
   3. Saliva; 4. Salivary amylase; 5. Lysozyme

Pharynx and Esophagus
1. Pharynx; 2. Reflex; 3. Bolus; 4. Muscles; 5. Soft
palate; 6. Epiglottis; 7. Peristaltic; 8. Esophageal
sphincter; 9. Heartburn

Stomach
A. 1. Fundus; 2. Body; 3. Pylorus
B. 1. Esophageal sphincter; 2. Pylorus; 3. Pyloric
   sphincter; 4. Rugae; 5. Body; 6. Fundus
C. 1. Smooth muscle; 2. Rugae; 3. Mucus; 4. Gastric
   juice; 5. Pepsin; 6. Hydrochloric acid; 7. Intrinsic
   factor; 8. Storage; 9. Chyme

Small Intestine
A. 1. Circular folds; 2. Villi; 3. Lacteals; 4. Microvilli;
   5. Ileocecal sphincter; 6. Ileocecal valve
B. 1. Duodenum; 2. Absorption; 3. Pancreas;
   4. Enzymes; 5. Microvilli; 6. Mucus; 7. Chyme;
   8. Smooth muscle

Large Intestine
A. 1. Cecum; 2. Appendix; 3. Colon; 4. Rectum;
   5. Anal canal; 6. Anus; 7. Internal anal sphincter;
   8. External anal sphincter
B. 1. Feces; 2. Water; 3. Mucus; 4. Microorganisms;
   5. Reproduce; 6. Vitamin K; 7. Defecation

Liver and Gallbladder
A. 1. Hepatic artery; 2. Hepatic portal vein;
   3. Hepatic veins; 4. Common hepatic duct;
   5. Cystic duct; 6. Common bile duct
B. 1. Absorbed materials; 2. Detoxify; 3. Bile;
   4. Bicarbonate ions; 5. Bile salts; 6. Bilirubin

Pancreas
1. Endocrine; 2. Insulin; 3. Glucose; 4. Exocrine;
5. Enzymes; 6. Bicarbonate ion; 7. Pancreatic duct

Peritoneum
1. Visceral peritoneum; 2. Parietal peritoneum;
3. Mesenteries; 4. Lesser omentum; 5. Greater
omentum; 6. Retroperitoneal

Regulation of the Digestive System
1. Medulla oblongata; 2. Parasympathetic;
3. Sympathetic; 4. CNS Reflexes; 5. Local reflexes;
6. Intramural plexus; 7. Endocrine cells; 8. Gastrin

Understanding Defecation
1. Rectum; 2. Abdominal muscles; 3. Mass
movements; 4. Intramural plexus; 5. Defecation
reflex; 6. External anal sphincter

Digestion and Absorption
A. 1. Digestion; 2. Absorption
B. 1. Polysaccharides; 2. Disaccharides;
   3. Monosaccharides; 4. Amylases;
   5. Monosaccharides
C. 1. Triglycerides; 2. Emulsification; 3. Bile salts;
   4. Lipase; 5. Micelles; 6. Lacteals; 7. Chyle
D. 1. Low-density lipoproteins (LDLs); 2. High-
   density lipoproteins (HDLs); 3. Low-density
   lipoproteins (LDLs); 4. High-density lipoproteins
   (HDLs)
E. 1. Pepsin; 2. Trypsin; 3. Amino acids
F. 1. Active transport; 2. Hepatic portal system;
   3. Modified; 4. Proteins; 5. Energy; 6. Fat

Disorders of the Digestive Tract
A. 1. Dental caries; 2. Periodontal disease; 3. Hiatal
   hernia; 4. Vomiting; 5. Peptic ulcer;
   6. Maldigestion; 7. Malabsorption
B. 1. Cancer; 2. Colonoscopy; 3. Appendicitis;
   4. Constipation; 5. Diarrhea; 6. Dysentery;
   7. Hemorrhoids
C. 1. Hepatitis; 2. Cirrhosis; 3. Hepatitis A;
   4. Hepatitis B and Hepatitis C; 5. Hepatitis C;
   6. Gallstones
D. 1. Staphylococcal food poisoning;
   2. Salmonellosis; 3. Typhoid; 4. Cholera;
   5. Giardiasis
E. 1. Tapeworms; 2. Pinworms; 3. Hookworms;
   4. Ascariasis

1. Mucosa, submucosa, muscularis, and serosa
2. Parotid glands, submandibular glands, and sublingual glands; salivary amylase
3. Fundus, body, and pylorus
4. Mucus: protection of stomach lining; Pepsin: enzyme that begins protein digestion; Hydrochloric acid: provides the correct pH for pepsin; Intrinsic factor: increases absorption of vitamin B_{12}
5. Circular folds, villi, and microvilli
6. Esophageal sphincter, pyloric sphincter, ileocecal sphincter, internal anal sphincter, and external anal sphincter
7. Excretory functions (elimination of bilirubin), digestive functions (bicarbonate ions, bile salts), stores and processes nutrients, synthesizes new molecules, and detoxifies harmful substances
8. Autonomic nervous system, local reflexes, and hormones
9. Salivary glands
10. Pharynx
11. Esophagus
12. Liver
13. Transverse colon
14. Ascending colon
15. Small intestine
16. Cecum
17. Appendix
18. Anal canal
19. Rectum
20. Sigmoid colon
21. Descending colon
22. Stomach

MASTERY LEARNING ACTIVITY

1. A   From the inner lining of the digestive tract to the outer layers are the mucosa, submucosa, muscularis, and serosa layers.

2. C   The innermost layer is the mucosa, containing simple or stratified epithelium; the next layer is the submucosa, a thick connective tissue layer containing blood vessels, nerves, and glands; the muscularis layer is next, and has both circular and longitudinal layers; the outermost layer is the serosa, which is composed of a serous membrane.

3. E   The tongue moves food about and helps (with the lips and cheeks) to hold the food between the teeth during mastication. The tongue also aids in swallowing, is involved with speech, and contains taste buds, which are involved with the sense of taste.

4. B   Each fourth of the adult mouth has two incisors, one canine, two premolars, and three molars (if the "wisdom teeth" are present).

5. D   Dentin is living, calcified, cellular material that surrounds the pulp cavity, and lies beneath the enamel found on the crown. Periodontal ligaments hold the teeth in place.

6. D   Parotid glands, submandibular glands, and sublingual glands are all pairs of salivary glands.

7. E   All of the listed processes are involved in swallowing. Pushing a bolus of food into the pharynx is voluntary; all the other processes listed are involuntary.

8. E   Bicarbonate ions are found in bile and pancreatic secretions. Bicarbonate ions neutralize acids by combining with hydrogen ions. As hydrogen ions decrease, pH increases, which would not allow pepsin to work efficiently.

9. B   The stomach, unlike most of the GI tract, has three layers of smooth muscle. The stomach connects to the small intestine at the pylorus, mucosal folds are called rugae, and chyme is the semifluid mixture in the stomach.

10. D   The stomach is protected from digestive enzymes and hydrochloric acid by the large amount of mucus that is secreted by the epithelial cells of the stomach lining.

11. D   Some contractions of the stomach function to mix gastric juice and the food within it to form chyme. Other contractions force chyme through the pyloric sphincter and into the small intestine. Gastric glands secrete gastric juice.

12. D   All of the structures listed increase the mucosal surface area of the small intestine.

13. B   Food passes from the stomach into the duodenum, then into the jejunum, into the ileum, and then into the large intestine.

14. A The small intestine, particularly the duodenum and jejunum, is the primary site for digestion and absorption in the GI tract. The ileocecal sphincter is located at the junction of the small intestine and large intestine. The epithelium lining the small intestine produces digestive enzymes, and the duodenum is the shortest portion of the three parts of the small intestine.

15. D Blood from the hepatic artery and hepatic portal vein bring blood into the liver. The hepatic veins exit the liver.

16. E The liver functions to produce bile, which is important in fat emulsification in the small intestine, it eliminates bilirubin in the bile, and it detoxifies harmful substances in the blood. Therefore all of the conditions listed could occur if the liver were inflamed and its functioning impaired.

17. B The exocrine gland secretion from the pancreas contains bicarbonate ions, which neutralize stomach acid, and digestive enzymes. The endocrine gland secretions, insulin and glucagon, are produced in the pancreatic islets (islets of Langerhans).

18. A Digestion begin in the oral cavity. Salivary amylase begins the process of carbohydrate digestion.

19. B Pepsin, found in the stomach, is the enzyme that begins the digestion of proteins. All of the other secretions are correctly matched with the location where they are produced.

20. E Stretch of the rectum initiates local and parasympathetic reflexes that result in defecation. Often the stretch is caused by the movement of feces into the rectum as a result of mass movements.

 FINAL CHALLENGES

1. Salivary amylase digests the starch in the bread, breaking the starch down into sugars. These sugars are responsible for the sweet taste.

2. Following a meal, stretch of the stomach initiates local reflexes that cause smooth muscle in the colon to contract. These strong peristaltic contractions, called mass movements, propel the contents of the colon into the rectum. Stretch of the rectum initiates the defecation reflex. Mass movements are most common about 15 minutes after breakfast.

3. Because the gallbladder provides storage for bile, removal of the gallbladder results in a relatively small amount of bile available at any one time. Bile salts are important for emulsifying fats. Therefore, a high-fat meal such as a hamburger and fries creates a difficult situation because only small amounts of bile are available for digestion of the fat. Large amounts of undigested fats in the intestine can produce gastric distress, possibly including bloating, cramping, and diarrhea.

4. Because enzymes are proteins, pepsin in the stomach and trypsin and peptidase in the small intestine digest the enzymes (i. e., break them down into amino acids), and the enzymes are no longer functional. Your best advice might be for your friend to save his money.

5. Cutting the vagus nerves is an older therapy for treatment of ulcers; newer therapy involves antibiotics because a specific bacterium has been implicated in many cases of peptic ulcer. Cutting the vagus nerves eliminates parasympathetic stimulation of the stomach, which reduces stomach acid secretion. The reduction in hydrochloric acid secretion could help in the treatment of a peptic ulcer. However, after a few months acid secretion often increases, and an ulcer may re-develop. Because elimination of parasympathetic stimulation also decreases stomach movement, food might be retained in the stomach. In some cases the stomach never really empties.

Nutrition and Metabolism

FOCUS: Metabolism is the total of all the chemical changes that occur in the body. Nutrients are the chemicals taken into the body that are used to produce energy, provide building blocks for new molecules, or function in other chemical reactions. The six major classes of nutrients are carbohydrates, proteins, lipids, vitamins, minerals, and water. Glucose is broken down in aerobic respiration to produce 38 ATP molecules or in anaerobic respiration to produce 2 ATP molecules. Fats and proteins can also be used to produce ATP. The energy in ATP is used for basal metabolism, physical activity and assimilation of food. As a byproduct of the chemical reactions of metabolism, heat is produced that contributes to body temperature. Heat can be exchanged with the environment by radiation, convection, conduction, and evaporation.

CONTENT LEARNING ACTIVITY

Introduction

❝Nutrition is the process by which food items are obtained and used by the body.❞

Match these terms with the correct statement or definition:

Anabolism
Catabolism

Metabolism
Nutrition

_____ 1. Includes digestion, absorption, transport, and cell metabolism.

_____ 2. Total of all the chemical changes that occur in the body.

_____ 3. Energy-requiring process by which small molecules are joined to form larger molecules.

_____ 4. Energy-releasing process by which large molecules are broken down into smaller molecules.

_____ 5. Energy released by this process is used to produce ATP.

Nutrients and Calories

66 *The six major nutrients are carbohydrates, proteins, lipids, vitamins, minerals, and water.* 99

Match these terms with the correct statement or definition:

Calorie   Fat
Carbohydrate   Protein
Essential nutrients

_____ 1. Minimum amounts of these (e.g., certain amino acids, linoleic acid, most vitamins, minerals, water, and carbohydrates) are required.

_____ 2. Amount of energy required to raise the temperature of 1000 g of water from 14° C to 15° C; used to express the amount of energy in food.

_____ 3. Contains approximately 9 Cal per gram.

Carbohydrates

66 *Carbohydrates include monosaccharides, disaccharides, and polysaccharides.* 99

Match these terms with the correct statement or definition:

Cellulose   Maltose
Fructose   Starch
Glucose   Sucrose
Glycogen

_____ 1. Monosaccharide converted into glucose by the liver.

_____ 2. Primary energy source for most cells.

_____ 3. Table sugar; a disaccharide of glucose and fructose.

_____ 4. Energy storage molecule produced from glucose in animals.

_____ 5. Not digestible by humans; provides "roughage."

Lipids

66 *Approximately 95% of the lipids in the human diet are triglycerides.* 99

Match these terms with the correct statement or definition:

Adipose   Saturated fat
Cholesterol   Triglycerides
Phospholipid   Unsaturated fat

_____ 1. Have only single covalent bonds between their carbon atoms.

_____ 2. Have one or more double covalent bonds between carbon atoms.

_____ 3. Excess triglycerides are stored in this tissue.

_____ 4. Part of cell membrane, modified to form bile salts and steroid hormones.

Proteins

"_Proteins are chains of amino acids._**"**

Match these terms with the correct statement or definition:

Antibodies   Essential
Collagen   Hemoglobin
Complete   Incomplete
Enzymes   Nonessential

_____ 1. Type of amino acids that can be manufactured by the body.

_____ 2. Food containing all the essential amino acids in the correct proportions.

_____ 3. Provides structural strength.

_____ 4. Regulate the rate of chemical reactions.

_____ 5. Transports oxygen and carbon dioxide.

_____ 6. Provide protection against microorganisms.

Sources and Recommended Requirements

"_A variety of foods provide the recommended requirement of carbohydrates, lipids, and proteins._**"**

A. Match these terms with the correct statement or definition:

Carbohydrate   Fat (saturated)
Cholesterol   Fat (unsaturated)
Fat (all)   Protein

_____ 1. Approximately 125 to 175 g are needed every day; otherwise acidosis or breakdown of muscle tissue occurs.

_____ 2. Should account for 30% or less of the total Caloric intake.

_____ 3. Should contribute no more than 10% of total fat intake.

_____ 4. Should be limited to 250 mg or less per day.

B. Match these terms with the
correct statement or definition:

Carbohydrate   Protein (complete)
Cholesterol   Protein (incomplete)
Monounsaturated fat   Saturated fat
Polyunsaturated fat

_____ 1. Fructose (in fruit), maltose (in cereal), lactose (in milk).

_____ 2. Lipid in fats of meat, whole milk, cheese, butter, coconut oil, and palm oil.

_____ 3. Lipid in olive and peanut oil.

_____ 4. Lipid in fish, safflower, sunflower, and corn oils.

_____ 5. Lipid in high concentration in brain, liver, and egg yolks.

_____ 6. Proteins in meat, fish, poultry, milk, cheese, and eggs.

_____ 7. Proteins in leafy green vegetables, grains, peas, and beans.

Vitamins

"*Vitamins are compounds that exist in minute quantities in food and are essential to normal metabolism.***"**

A. Match these terms with the
correct statement or definition:

Fat-soluble vitamins
Water-soluble vitamins

_____ 1. B-complex vitamins and vitamin C.

_____ 2. Vitamins A, D, E, and K.

_____ 3. Vitamins that can be stored in the body.

B. Match these vitamins with the
correct deficiency symptom:

A (retinol)   D (cholecalciferol)
B_{12} (cobalamin)   K (phylloquinone)
C (ascorbic acid)

_____ 1. Scurvy.

_____ 2. Night blindness, retarded growth, and skin disorders.

_____ 3. Excessive bleeding resulting from retarded blood clotting.

_____ 4. Pernicious anemia and nervous system disorders.

_____ 5. Rickets.

Minerals

"_Minerals are inorganic nutrients necessary for normal metabolic functions._**"**

Match these minerals with
the correct function:

Calcium   Phosphorus
Iodine   Potassium
Iron   Sodium

_____ 1. Bone and teeth formation, blood clotting, muscle activity, and nerve
function.

_____ 2. Regulation of water balance (osmosis); nerve and muscle function.

_____ 3. Important part of ATP.

_____ 4. Component of hemoglobin that binds with oxygen.

_____ 5. Thyroid hormone production; maintenance of normal metabolic rate.

Cell Metabolism

"_The chemical processes that occur within cells are often referred to as cell metabolism._**"**

A. Match these terms with the
correct statement or definition:

Aerobic respiration   Glucose
Anaerobic respiration   High energy electrons
ATP   Oxygen
Carbohydrate pathway   Pyruvic acid
Carbon dioxide   Water

_____ 1. Energy currency of the cell; releases energy used to drive chemical
reactions.

_____ 2. Two of these molecules are formed from glucose in the first phase of
anaerobic and aerobic respiration.

_____ 3. Breakdown of glucose to lactic acid molecule in the absence of oxygen;
produces two ATP molecules.

_____ 4. Breakdown of glucose to carbon dioxide and water in the presence of
oxygen; produces 38 ATP molecules.

_____ 5. Molecule formed using the carbon atoms in pyruvic acid.

_____ 6. Contain energy derived from pyruvic acid; used to produce most of the
ATP in aerobic respiration.

_____ 7. Molecule used in the last step of aerobic respiration to form water.

_____ 8. "Backbone" of cellular metabolism: excess fats and proteins can be
used for energy, and carbohydrates can be used to manufacture fats
and amino acids.

B. Match these terms with
   the correct terms labeled
   in Figure 17-1:

Aerobic respiration   Lactic acid
Anaerobic respiration   Pyruvic acid
Carbon dioxide   Water
High energy electrons

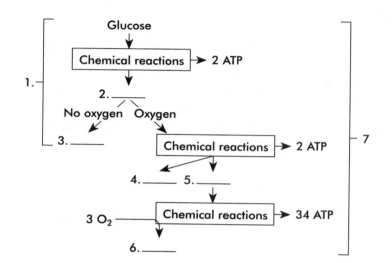

1. _____

2. _____

3. _____

4. _____

5. _____

6. _____

7. _____

Understanding Starvation and Obesity

66 Starvation and obesity can be understood in term of metabolism. 99

A. Match these terms with the
   correct statement or definition:

First
Second
Third

_____ 1. Phase of starvation in which blood glucose levels are maintained
                              through the metabolism of glycogen, fats, and proteins.

_____ 2. Phase of starvation in which fats are the primary energy source: the
                              brain uses ketones for energy.

_____ 3. Phase of starvation in which proteins are the primary energy source.

B. Match these terms with the
   correct statement or definition:

Decreased   Increased
Hyperplastic obesity Set point
Hypertrophic obesity

_____ 1. There is a greater than normal number of fat cells that are also larger
                              than normal.

_____ 2. There is a normal number of fats cells that have increased in size.

_____ 3. Most common type of obesity; associated with moderate overweight.

_____ 4. Term for the amount of fat maintained by the body.

_____ 5. Effect on appetite of below normal weight.

_____ 6. Effect on basal metabolic rate of below normal weight.

Metabolic Rate

"*The metabolic rate is the total amount of energy produced and used by the body per unit of time.***"**

Match these terms with the correct statement or definition:

Assimilation
Basal metabolic rate

Caloric intake
Skeletal muscle contractions

_____ 1. The minimum energy needed to keep the resting body functional; usually determined by measuring oxygen consumption.

_____ 2. The energy needed to produce digestive enzymes and absorb foods.

_____ 3. Accounts for approximately 60% of total energy expenditure.

_____ 4. Accounts for approximately 10% of total energy expenditure.

_____ 5. Reducing this factor can result in weight loss.

_____ 6. Increasing this factor can result in weight loss.

Body Temperature

"*Humans can maintain a constant body temperature even though environmental temperatures vary.***"**

Match these terms with the correct statement or definition:

Conduction
Convection
Evaporation
Heat
Hypothalamus

Radiation
Set point
Vasoconstriction
Vasodilation

_____ 1. Produced as a byproduct of the chemical reactions of metabolism.

_____ 2. Transfer of heat between the body and the air.

_____ 3. Loss or gain of heat from the body by infrared energy.

_____ 4. Increases heat loss by changing blood flow through the skin.

_____ 5. Part of the brain that controls the body's heat exchange mechanisms.

_____ 6. Temperature which the body maintains through negative-feedback mechanisms.

Hyperthermia and Hypothermia

"*Excessive heat gain or loss can produce abnormal body temperatures.***"**

Match these terms with the correct statement or definition:

Fever   Heat stroke
Frostbite   Hypothermia
Heat exhaustion   Hyperthermia

_____ 1. General term for higher than normal body temperature; occurs when heat gain exceeds the ability of the body to lose heat.

_____ 2. Negative-feedback mechanisms are working, but are unable to prevent hyperthermia; skin is cool and wet.

_____ 3. Hyperthermia that results from a breakdown of negative-feedback mechanisms; skin is dry and flushed.

_____ 4. Results when pyrogens raise the temperature set point of the hypothalamus.

_____ 5. Symptoms include shivering, sluggish thinking, and uncoordinated movements.

_____ 6. Damage to the skin resulting from prolonged exposure to the cold.

QUICK RECALL

1. List the six major classes of nutrients.

2. Give two examples of monosaccharides, disaccharides, and complex carbohydrates.

3. State the difference between saturated fats and unsaturated fats; between essential amino acids and nonessential amino acids; between a complete protein food and an incomplete protein food.

4. List the fat soluble vitamins.

5. List the molecules and the number of ATP produced by anaerobic and aerobic respiration.

6. List the three ways metabolic energy can be used.

7. List four ways heat is exchanged between the body and the external environment.

MASTERY LEARNING ACTIVITY

Place the letter corresponding to the correct answer in the space provided.

_____ 1. Which of the following statements concerning Calories is true?
a. A Calorie is a measure of heat.
b. There are 4 Cal in a gram of protein.
c. A pound of body fat has 3500 Cal.
d. all of the above

_____ 2. Complex carbohydrates
a. include sucrose.
b. can be found in large amounts in milk.
c. form energy storage molecules in plants and animals.
d. all of the above

_____ 3. A good source of monounsaturated fats is
a. the fats of meats.
b. egg yolks.
c. whole milk.
d. fish oil.
e. olive oil.

_____ 4. Triglycerides
a. are an important source of energy for the brain.
b. are modified to form bile salts and steroid hormones
c. are part of the plasma membrane and myelin sheaths
d. in excess are stored in adipose tissue.

_____ 5. A complete protein food
a. provides the daily amount (grams) of protein recommended in a healthy diet.
b. can be used to synthesize the nonessential amino acids.
c. contains all 20 amino acids.
d. includes beans, peas, and leafy green vegetables.

_____ 6. Proteins
a. regulate the rate of chemical reactions.
b. function as carrier molecules.
c. provide protection against microorganisms.
d. all of the above

_____ 7. Concerning vitamins,
a. most can be synthesized by the body.
b. they are normally broken down before they can be used by the body.
c. A, D, E, and K are water-soluble vitamins.
d. they function with or as part of enzymes.

8. Minerals
   a. are inorganic nutrients.
   b. compose approximately 4% of total body weight.
   c. function with or as part of enzymes.
   d. all of the above

9. Anaerobic respiration occurs in the _____ of oxygen and produces _____ energy for the cell than does aerobic respiration.
   a. absence, less
   b. absence, more
   c. presence, less
   d. presence, more

10. Which of the following are produced in both anaerobic and aerobic respiration?
    a. lactic acid
    b. pyruvic acid
    c. carbon dioxide
    d. water

11. The production of most ATP in aerobic respiration requires high energy electrons and
    a. alcohol.
    b. water.
    c. oxygen.
    d. lactic acid.

12. The carbon dioxide you breathe out comes from
    a. the breakdown of lactic acid.
    b. the breakdown of water.
    c. anaerobic respiration.
    d. the food you eat.

13. Which of the following can be used as a source of energy to make ATP?
    a. carbohydrates
    b. fats
    c. proteins
    d. all of the above

14. The major use of energy by the body is for
    a. basal metabolism.
    b. physical activity.
    c. assimilation of food.
    d. thinking of answers to these questions.

15. The loss of heat caused by the loss of water from the surface of the body is called
    a. radiation.
    b. evaporation.
    c. conduction.
    d. convection.

☆ —— FINAL CHALLENGES —— ☆

Use a separate sheet of paper to complete this section.

1. When a person is trying to lose weight, a reduction in caloric input and exercise is recommended. Give three reasons why exercise would help to reduce weight.

2. It is recommended that a person on a diet drink six to eight glasses of cool water per day. How could this practice help a person to lose weight?

3. Suppose a typical male and female of the same weight went on a diet to lose weight. If they both ate the same amount and kind of food and did the same amount and kind of exercise, would they both lose weight at the same rate?

4. Explain how increased physical activity increases metabolic rate and oxygen consumption during exercise.

5. A woman is jogging on a warm day. The air temperature is 100° F. Describe the ways in which she gains and loses heat.

ANSWERS TO CHAPTER 17

Introduction
1. Nutrition; 2. Metabolism; 3. Anabolism;
4. Catabolism; 5. Catabolism;

Nutrients and Calories
1. Essential nutrients; 2. Calorie; 3. Fat

Carbohydrates
1. Fructose; 2. Glucose; 3. Sucrose; 4. Glycogen;
5. Cellulose

Lipids
1. Saturated fat; 2. Unsaturated fat; 3. Adipose;
4. Cholesterol

Proteins
1. Nonessential; 2. Complete; 3. Collagen;
4. Enzymes; 5. Hemoglobin; 6. Antibodies

Sources and Recommended Requirements
A. 1. Carbohydrate; 2. Fat (all); 3. Fat (saturated);
4. Cholesterol
B. 1. Carbohydrate; 2. Saturated fat;
3. Monounsaturated fat; 4. Polyunsaturated fat;
5. Cholesterol; 6. Protein (complete); 7. Protein
(incomplete)

Vitamins
A. 1. Water-soluble vitamins; 2. Fat-soluble
vitamins; 3. Fat-soluble vitamins
B. C (ascorbic acid); 2. A (retinol);
3. K (phylloquinone); 4. B_{12} (cobalamin);
5. D (cholecalciferol)

Minerals
1. Calcium; 2. Sodium; 3. Phosphorus; 4. Iron;
5. Iodine

Cell Metabolism
A. 1. ATP; 2. Pyruvic acid; 3. Anaerobic respiration;
4. Aerobic respiration; 5. Carbon dioxide;
6. High energy electrons; 7. Oxygen;
8. Carbohydrate pathway
B. 1. Anaerobic respiration; 2. Pyruvic acid;
3. Lactic acid; 4. Carbon dioxide; 5. High energy
electrons; 6. Water; 7. Aerobic respiration

Understanding Starvation and Obesity
A. 1. First; 2. Second; 3. Third
B. Hyperplastic obesity; 2. Hypertrophic obesity;
3. Hypertrophic obesity; 4. Set point;
5. Increased; 6. Decreased

Metabolic Rate
1. Basal metabolic rate; 2. Assimilation; 3. Basal
metabolic rate; 4. Assimilation; 5. Caloric intake;
6. Skeletal muscle contractions

Body Temperature
1. Heat; 2. Convection; 3. Radiation; 4. Vasodilation;
5. Hypothalamus; 6. Set point

Hyperthermia and Hypothermia
1. Hyperthermia; 2. Heat exhaustion; 3. Heat stroke;
4. Fever; 5. Hypothermia; 6. Frostbite

1. Carbohydrates, lipids, proteins, vitamins, minerals, and water
2. Monosaccharides: glucose and fructose; disaccharides: sucrose, maltose, and lactose; complex carbohydrate include the polysaccharides starch, glycogen, and cellulose
3. Saturated fats have only one covalent bond between carbon atoms, whereas unsaturated fats have double covalent bonds; essential amino acids must be ingested, whereas nonessential amino acids can be synthesized by the body; a complete protein food has all eight essential amino acids in the correct proportions, whereas an incomplete protein food does not
4. Vitamins, A, D, E, and K
5. Anaerobic respiration: lactic acid and 2 ATP; aerobic respiration; carbon dioxide, water, and 38 ATP
6. Basal metabolism, muscle contraction, and assimilation of food
7. Radiation, convection, conduction, and evaporation

1. D. A Calorie is the amount of heat required to raise the temperature of 1000 g of water from 14° to 15° C. There are 4 Cal in a gram of protein or carbohydrate and 9 Cal in a gram of fat. A pound of body fat contains 3500 Cal.

2. C. Complex carbohydrates are large polysaccharides and include starch (energy storage in plants), glycogen (energy storage in animals), and cellulose (roughage). Sucrose (table sugar) and lactose (milk sugar) are disaccharides.

3. E. Olive and peanut oils are good sources of monounsaturated fats. Polyunsaturated fats are in fish, safflower, sunflower, and corn oils. Saturated fats are in the fats of meats, whole milk, cheese, butter, eggs, nuts, coconut oil, and palm oil. Egg yolks are high in cholesterol.

4. D. Triglycerides are 95% of the lipids ingested by humans. Excess triglycerides are stored in adipose tissue. Glucose is an important source of energy for the brain, cholesterol is modified to form bile salts and steroid hormones, and phospholipids are part of the plasma membrane and myelin sheaths.

5. B. A complete protein food contains the eight essential amino acids, from which the nonessential amino acids can be synthesized. Complete protein foods include meat, fish, poultry, milk, cheese, and eggs.

6. D. Proteins include enzymes (regulate the rate of chemical reactions), carrier molecules, and antibodies (provide protection against microorganisms).

7. D. Vitamins function with or as parts of enzymes. Most vitamins, which are called essential vitamins, cannot be synthesized in the body. Vitamins are not broken down before they are used. Vitamins A, D, E, and K are fat-soluble vitamins.

8. D. Minerals are inorganic nutrients that are necessary for normal metabolic functions.

9. A. Anaerobic respiration does not require oxygen and produces fewer ATP than aerobic respiration.

10. B. Glucose is broken down to pyruvic acid in the first phase of anaerobic and aerobic respiration. Lactic acid is produced in anaerobic respiration, and carbon dioxide and water are produced in aerobic respiration.

11. C. The breakdown of pyruvic acid results in the transfer of energy from the chemical bonds of pyruvic acid to high energy electrons. This energy is used to produce ATP. The last step in this process is the combination of oxygen with hydrogen to form water.

12. D. The food you eat contains organic molecules such as glucose. The chemical bonds between the carbon atoms of these molecules is broken down during aerobic respiration to release energy used to produce ATP. The carbon atoms combine with oxygen to from carbon dioxide, which is breathed out.

13. D. Carbohydrates, fats, and proteins can all enter the carbohydrate pathway and be used to produce ATP.

14. A. Basal metabolism accounts for 60% of energy expenditure, muscular activity 30%, and assimilation of food 10%.

15. B. The evaporation of water from the surface of the skin eliminates heat.

 FINAL CHALLENGES

1. First, exercise increases energy (Calories) usage. Second, basal metabolic rate is elevated following exercise because of increased body temperature (as in fever). Third, exercise increases the proportion of muscle tissue to adipose tissue in the body. Because muscle tissue is metabolically more active than adipose tissue, basal metabolic rate increases. Fourth, during exercise epinephrine levels increase, resulting in increased blood sugar levels that can depress the hunger center in the brain and reduce food consumption immediately following exercise.

2. Drinking cool water could help in two ways. Because the water is cool, raising the water to body temperature requires the expenditure of Calories. Also, stretch of the stomach decreases appetite.

3. All else being equal, the typical male loses weight more rapidly because males have a higher basal metabolic rate that females.

4. Metabolic rate is the total amount of energy produced and used by the body per unit of time. Increased physical activity results from muscle contractions, which require energy in the form of ATP. Thus the body produces and uses more energy during exercise and metabolic rate increases. The energy in ATP is mostly derived from aerobic respiration, which requires oxygen. Metabolic rate is typically determined by measuring oxygen consumption. An increase in oxygen consumption indicates an increase in metabolic rate because the body is producing more energy through aerobic respiration.

5. The primary sources of heat are the heat she produces as a byproduct of her own metabolism and heat gained from the environment. Environmental heat gain could include radiation from the sun and surrounding objects, and convective heat gain from the warm air. There is probably little heat gain by conduction through her feet. Heat loss is accomplished by evaporation.

Urinary System and Fluid Balance

FOCUS: The urinary system consists of the kidneys, ureters, urinary bladder and urethra. The kidneys remove waste products from the blood; help control the volume, ion concentration, and pH of the blood; and control red blood cell production. The nephron, which is the functional unit of the kidney, utilizes filtration, reabsorption, and secretion in the formation of urine. Of the filtrate formed in the nephron, 99% is reabsorbed. Blood concentration, blood volume, and hormones regulate urine production. Most body fluids are inside cells, but water and electrolytes move between intracellular and extracellular compartments. Regulation of fluid and electrolyte balance is necessary to maintain homeostasis. The mechanisms that regulate body pH are critical for survival.

CONTENT LEARNING ACTIVITY

Kidneys

66 *The kidneys are bean-shaped organs, and each is about the size of a tightly-clenched fist.* **99**

A. Match these terms with the correct statement or definition:

Hilum
Calyces
Renal pelvis

Renal sinus
Ureter

_____ 1. Location on the medial side of the kidney where nerves, the renal artery and vein, and the ureter attach to the kidney.

_____ 2. Cavity near the hilum; filled with fat and connective tissue.

_____ 3. Enlarged urinary channel in the center of the renal sinus.

_____ 4. Funnel-shaped structures; extend from renal pelvis to kidney tissue.

_____ 5. Tube that carries urine; formed as the renal pelvis narrows.

B. **M**atch these terms with the correct statement or definition:

Cortex   Renal pyramids
Medulla

_____

1. Outer portion of the kidney.

_____

2. Inner portion of the kidney, containing the renal pyramids.

_____

3. Cone-shaped structures found in the medulla of the kidney.

C. **M**atch these terms with the correct parts labeled in Figure 18-1:

Calyx
Connective tissue capsule
Cortex
Medulla
Renal pelvis
Renal pyramid
Ureter

1. _____

2. _____

3. _____

4. _____

5. _____

6. _____

7. _____

Renal artery

Hilum (indentation)

Renal vein

The Nephron

The functional unit of the kidney is the nephron.

A. Match these terms with the correct statement or definition:

Ascending limb
Collecting duct
Descending limb
Distal convoluted tubule

Filtrate
Proximal convoluted tubule
Urine

_____ 1. Fluid filtered from the glomerulus into Bowman's capsule.

_____ 2. From Bowman's capsule, filtrate passes into this part of the nephron.

_____ 3. Portion of the loop of Henle that extends back toward the cortex.

_____ 4. Filtrate passes from the loop of Henle into this part of the nephron.

_____ 5. Tubule that extends from the cortex through the medulla and empties into a calyx; connected to the distal convoluted tubule.

_____ 6. Fluid produced from filtrate by nephrons and collecting ducts.

B. Match these terms with the correct parts labeled in Figure 18-2:

Afferent arteriole
Bowman's capsule
Collecting duct
Distal convoluted tubule
Efferent arteriole
Glomerulus
Loop of Henle
Peritubular capillaries
Proximal convoluted tubule

1. _____

2. _____

3. _____

4. _____

5. _____

6. _____

7. _____

8. _____

9. _____

DESNOYER ©

Arteries and Veins

"Approximately 20% of the blood pumped by the heart each minute flows through the kidneys."

Match these terms with the
correct statement or definition:

Afferent arteriole   Peritubular capillaries
Efferent arteriole   Renal arteries

_____ 1. Vessels that branch off the abdominal aorta and enter each kidney
                            at the hilum.

_____ 2. Vessel that carries blood to each glomerulus.

_____ 3. Vessel that carries blood away from each glomerulus.

_____ 4. Small vessels extending from efferent arterioles; surround proximal
                            and distal convoluted tubules and loop of Henle.

Ureters, Urinary Bladder, and Urethra

"The ureters and urethra are small tubes that carry urine, and the bladder is a hollow muscular structure that stores urine.

Match these terms with the
correct statement or definition:

External urinary sphincter   Ureters
Internal urinary sphincter   Urethra
Smooth muscle   Urinary bladder

_____ 1. Small tubes that carry urine from the renal pelvis to the urinary
                            bladder.

_____ 2. A hollow, muscular container in the pelvic cavity; stores urine.

_____ 3. Tube that exits the urinary bladder.

_____ 4. Cells that form the walls of the ureters and urinary bladder.

_____ 5. Skeletal muscle that surrounds the urethra near the urinary bladder.

Urine Production

"The three processes critical to the formation of urine are filtration, reabsorption, and secretion."

A. Match these terms with the
   correct statement or definition:

Ammonia
Urea

_____ 1. Toxic substance produced when amino acids are used as a source of
                            energy.

_____ 2. Substance produced in the liver and carried by the blood to the
                            kidneys, where it is eliminated.

B. Match these terms with the correct statement or definition:

Filtration   Tubular secretion
Tubular reabsorption

_____ 1. Movement of fluid from the blood across the wall of the glomerulus and into Bowman's capsule of the nephron.

_____ 2. Movement of substances from the filtrate into the peritubular space and back into the blood of the peritubular capillaries.

_____ 3. Movement of substances across the nephron wall into the filtrate.

☞ Urine produced by the nephrons consists of the substances that are filtered and secreted into the nephron minus those substances that are reabsorbed.

C. Using the terms provided, complete the following statements:

Blood cells   Filtration pressure
Decreases   Increases
Filtrate   Water
Filtration membrane

Together, the wall of the glomerulus and the wall of Bowman's capsule are called the _(1)_; the portion of the plasma crossing the filtration membrane to enter Bowman's capsule becomes the _(2)_. The filtration membrane prevents the entry of _(3)_ and proteins into Bowman's capsule, but _(4)_ and ions and molecules of small diameter readily pass through the filtration barrier. The formation of filtrate depends on a pressure difference called _(5)_, between the glomerulus and Bowman's capsule. When filtration pressure increases, the volume of the filtrate _(6)_, and the urine volume _(7)_. Filtration pressure _(8)_ when sympathetic stimulation produces constriction of the afferent arteriole during excitement or vigorous physical activity.

1. _____
2. _____
3. _____
4. _____
5. _____
6. _____
7. _____
8. _____

D. Using the terms provided, complete the following statements:

Osmosis   Renal veins
Peritubular space   Secreted
Proteins and glucose   Urine
Reabsorbed   Waste products

As the filtrate flows through the nephron and collecting ducts, many of the substances in the filtrate are _(1)_. Water moves by _(2)_ across the wall of the nephron with the ions and molecules that are reabsorbed and enters the _(3)_. The filtrate that enters the peritubular space then enters the peritubular capillaries and flows through the _(4)_ to enter the general circulation. Only about 1% of the filtrate becomes _(5)_, which contains a high concentration of _(6)_. Reabsorption prevents the loss of useful substances such as sodium and chloride ions and organic molecules such as _(7)_. However, ammonia, excess hydrogen ions, and excess potassium ions are _(8)_ across the wall of the nephron into the filtrate.

1. _____
2. _____
3. _____
4. _____
5. _____
6. _____
7. _____
8. _____

Regulation of Urine Concentration and Volume

"The volume and composition of urine changes, depending on conditions in the body."

A. **Complete each statement by providing the missing word:**

Increase(s)
Decrease(s)

_____ 1. If blood concentration increases, the amount of urine produced _____.

_____ 2. If blood concentration increases, the concentration of the urine _____.

_____ 3. A smaller volume of concentrated urine _____ water conservation.

_____ 4. If blood volume (or blood pressure) increases, urine volume _____.

_____ 5. Increased loss of water in the urine causes blood volume (and blood pressure) to _____.

B. **Complete each statement by providing the missing word:**

Increase(s)
Decrease(s)

_____ 1. ADH _____ the amount of water reabsorbed by the distal convoluted tubules and collecting ducts.

_____ 2. ADH _____ the volume of urine produced.

_____ 3. ADH _____ the concentration of the urine.

_____ 4. If the concentration of the blood increases, ADH secretion _____.

☞ Diabetes insipidus results when the posterior pituitary fails to secrete ADH, or when the kidney tubules cannot respond to ADH. People with diabetes insipidus produce a large volume of dilute urine each day.

C. **Complete each statement by providing the missing word:**

Decrease(s)
Increase(s)

_____ 1. Aldosterone _____ reabsorption of sodium and chloride ions from the nephron.

_____ 2. Decreased blood pressure causes renin production by the juxtaglomerular apparatus in the kidney to _____.

_____ 3. Renin, an enzyme, _____ production of angiotensin I, which is converted to angiotensin II.

_____ 4. Angiotensin II _____ secretion of aldosterone by the adrenal cortex.

_____ 5. Increased aldosterone secretion _____ urine volume.

_____ 6. Increased aldosterone secretion _____ blood volume.

_____ 7. Increased blood volume _____ blood pressure.

The juxtaglomerular apparatus is a structure formed where the specialized cells of the afferent arteriole and specialized cells of the distal convoluted tubule come into contact.

D. Complete each statement by providing the missing word:

Decrease(s)
Increase(s)

_____ 1. Increased stretch of the atria of the heart _____ secretion of atrial natriuretic hormone.

_____ 2. Atrial natriuretic factor _____ urine volume and salt concentration of the urine.

_____ 3. Atrial natriuretic factor _____ blood volume.

_____ 4. Diuretics _____ urine volume.

_____ 5. Alcohol _____ urine volume by inhibiting ADH production.

_____ 6. Caffeine and related substances increase urine volume because of a(n) _____ in filtration pressure.

Urine Movement

66 *The micturition reflex is initiated by stretching of the bladder wall.* **99**

Match these terms with the correct statement or definition:

Higher brain centers
Spinal cord reflex

_____ 1. Causes the urinary bladder to contract and the internal and external urinary sphincters to relax.

_____ 2. Can inhibit or stimulate the micturition reflex.

Irritation of the urinary bladder or urethra such as occurs during bacterial infections may initiate the urge to urinate, even though the bladder may be nearly empty.

Body Fluid Compartments

66 *Water and electrolytes move between two major compartments, but their movement is regulated.* **99**

Match these terms with the correct statement or definition:

Extracellular fluid compartment
Intracellular fluid compartment

_____ 1. Water and electrolytes inside cells; about 63% of total body water.

_____ 2. Fluid outside the cells (about 37% of total body water); includes interstitial fluid, plasma in blood vessels, and lymph in vessels.

_____ 3. Protein, and potassium, phosphate, magnesium, and sulfate ions, are present in greater concentration; sodium, chloride, and bicarbonate ions present in lesser concentration.

Water moves between the intracellular and extracellular fluid compartments continuously. Thus, changes in the concentration of ions and molecules results in movement of water by osmosis.

Regulation of Extracellular Fluid Composition

66*Homeostasis requires that the intake of water and electrolytes is equal to their elimination.*99

A. Complete each statement by providing the missing word:

Decrease(s)
Increase(s)

_____ 1. When concentration of the blood increases, thirst _____.

_____ 2. If blood pressure decreases, such as during shock, thirst _____.

_____ 3. Consumption of water _____ blood volume and blood pressure.

_____ 4. If the mouth becomes dry, thirst _____.

The most important organ regulating the loss of water and electrolytes from the body is the kidney.

B. Match these terms with the correct location on figure 18-3 by placing a D in the blank for those substances that are decreased, and an I for those substances that are increased.

Decreased
Increased

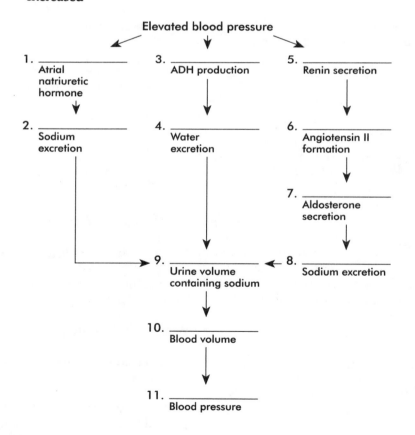

274

C. Using the terms provided, complete the following statements:

Aldosterone   Kidneys
Decrease(s)   Potassium
Increase(s)   Sodium

_(1)_ ions, which are the dominant extracellular ions, are
regulated by the same mechanisms that regulate blood pressure
and extracellular fluid concentration. The _(2)_ are the major
route by which excess sodium ions are excreted. Electrically
excitable tissue, such as nerve and muscle, are highly sensitive
to slight changes in the extracellular _(3)_ ion concentration.
The hormone _(4)_ plays a major role in regulating the
concentration of potassium ions in extracellular fluid. When
levels of potassium in the blood _(5)_, aldosterone secretion is
strongly stimulated. Aldosterone increases sodium ion
reabsorption, but it also _(6)_ potassium ion secretion into the
nephron, returning potassium ion concentration toward the
normal level.

1. _____

2. _____

3. _____

4. _____

5. _____

6. _____

Regulation of Acid-Base Balance

66*The mechanisms that regulate body pH are critical for survival.*99

A. Match these terms with the   Bicarbonate buffer system   Protein and phosphate
   correct statement or definition: buffer system

_____   1. Combine with the largest numbers of hydrogen ions.

_____   2. Can be regulated by the respiratory system and the kidneys.

B. Using the terms provided, complete the following statements:

Carbon dioxide   Increase(s)
Decrease(s)   Rapidly
Hydrogen ions   Slowly

The respiratory system responds _(1)_ to changes in pH. If
metabolic activity increases, or respiration is not adequate,
_(2)_ accumulates in the body fluids. As carbon dioxide levels
increase, pH of the body fluids _(3)_, neurons in the respiratory
center of the brain are stimulated, and the rate and depth of
respiration _(4)_. As a result, carbon dioxide elimination
_(5)_ and the concentration of carbon dioxide in the body _(6)_,
which causes pH to _(7)_ to its normal range. The kidneys
regulate pH, but react _(8)_ compared to the respiratory system.
Cells in the wall of the nephron can regulate pH directly by
secreting _(9)_ into the urine. If the pH of the body fluids
decreases below normal, the rate at which the kidneys secrete
hydrogen ions _(10)_. Consequently, the pH of the blood _(11)_
toward its normal value.

1. _____

2. _____

3. _____

4. _____

5. _____

6. _____

7. _____

8. _____

9. _____

10. _____

11. _____

Understanding Acidosis and Alkalosis

"*Failure to maintain normal pH levels can result in acidosis or alkalosis.*"

Using the terms provided, complete the following statements:

Acidosis
Alkalosis
Central nervous system
Hyperexcitability

Kidneys
Metabolic
Respiratory

When the pH value of the body fluids is below 7.35, the condition is referred to as (1) . When this occurs, the (2) malfunctions, and the individual becomes disoriented and possibly comatose. Acidosis is separated into two categories. (3) acidosis results when the respiratory system is unable to eliminate adequate carbon dioxide. On the other hand, (4) acidosis results from the excessive production of acidic substances because of increased metabolism, or decreased ability of the kidneys to eliminate hydrogen ions in the urine. When the pH of the body fluids is above 7.45, the condition is called (5) . A major effect of alkalosis is (6) of the nervous system, which can lead to spasms, convulsions, or even tetany of the respiratory muscles and death. (7) alkalosis results from hyperventilation, such as in response to stress. (8) alkalosis usually results from rapid elimination of hydrogen ions during severe vomiting, or from excess aldosterone secretion. In case of respiratory acidosis or alkalosis, the (9) help compensate for the change in pH; in the case of metabolic acidosis or alkalosis, the (10) system helps compensate for the change in pH.

1. _____
2. _____
3. _____
4. _____
5. _____
6. _____
7. _____
8. _____
9. _____
10. _____

Disorders of the Urinary System

"*Proper function of the kidneys is critical for survival.*"

A. Match these terms with the correct statement or definition:

Acute glomerular nephritis
Chronic glomerular nephritis

Cystitis
Kidney stones

_____ 1. Inflammation of the wall of the glomerulus and the wall of Bowman's capsule; occurs 1 to 3 weeks after severe bacterial infection; normally subsides after several days.

_____ 2. Inflammation of the wall of the glomerulus and the wall of Bowman's capsule that is long term and usually progressive.

_____ 3. Inflammation of the urinary bladder.

_____ 4. Precipitates of substances such as calcium salts that usually form in the renal pelvis.

B. Match these terms with the
correct statement or definition:

Acute renal failure   Dialysis
Chronic renal failure   Kidney transplant

_____ 1. Rapid and extensive damage to the kidney; can result from renal tubule blockage, acute glomerular nephritis, or toxic substances.

_____ 2. Permanent damage to nephrons that results in inadequate kidney function; leads to toxin accumulation, edema, elevated potassium levels, and acidosis.

_____ 3. A procedure that substitutes for the excretory functions of the kidney; an artificial selectively permeable membrane is used to remove metabolic wastes from the blood.

_____ 4. Use of a donor kidney to replace the kidney of a patient suffering from acute renal failure.

QUICK RECALL

1. List the five major parts of a nephron.

2. List the three steps in urine formation.

3. Name three hormones that affect urine production, and give the major effect of each on urine production.

4. Following stretch of the bladder, list the events that result in micturition.

5. List three ways the sensation of thirst is increased.

6. State the effect of ADH, aldosterone, and renin on urine production, blood volume, and blood pressure.

7. Name three important buffer systems in the body.

8. State the effect on blood pH when respiration rate increases above normal and decreases below normal.

MASTERY LEARNING ACTIVITY

Place the letter corresponding to the correct answer in the space provided.

_____ 1. Nephrons are found in the
a. renal cortex.
b. renal medulla.
c. renal pelvis.
d. hilum.
e. both a and b

_____ 2. Which of the following structures contain blood?
a. glomerulus
b. Bowman's capsule
c. collecting duct
d. loop of Henle
e. a and b

_____ 3. The functional unit of the kidney is the
a. renal pelvis.
b. renal sinus.
c. nephron.
d. calyx.

_____ 4. The tip of each renal pyramid is surrounded by
a. the glomerulus.
b. a calyx.
c. Bowman's capsule.
d. the loop of Henle.

_____ 5. Given the following structures:
1. Bowman's capsule
2. collecting duct
3. distal convoluted tubule
4. loop of Henle
5. proximal convoluted tubule

Choose the arrangement that lists the structures in order as filtrate leaves the glomerulus and travels to the renal calyx.
c. 1,3,4,5,2
a. 1,5,4,3,2
d. 2,5,3,4,1
b. 4,1,2,3,5

_____ 6. Kidney function is accomplished by which of the following mechanisms?
a. secretion
b. filtration
c. reabsorption
d. all of the above

_____ 7. Given the following vessels:
1. afferent arteriole
2. efferent arteriole
3. peritubular capillaries

Choose the path a red blood cell would take as it passed from the renal artery to the renal vein.
a. 1,2,3
b. 2,1,3
c. 2,3,1
d. 3,2,1

_____ 8. The urinary bladder
a. has walls composed of skeletal muscle.
b. produces urine.
c. is connected to the outside of the body by ureters.
d. is located in the pelvic cavity.

_____ 9. Which of the following reduces filtration pressure in the glomerulus?
a. elevated blood pressure
b. constriction of afferent arterioles
c. cardiovascular shock
d. b and c
e. all of the above

_____ 10. Which of the following is reabsorbed by the nephron?
a. glucose
b. water
c. sodium ions
d. chloride ions
e. all of the above

_____ 11. About ____% of filtrate volume is reabsorbed in the nephron.
a. 10
b. 50
c. 63
d. 75
e. 99

_____ 12. Which of the following is NOT secreted across the wall of the nephron into the filtrate?
a. excess sodium ions
b. excess potassium ions
c. excess hydrogen ions
d. ammonia

_____ 13. Increased ADH results in
a. decreased blood volume.
b. increased permeability of the distal convoluted tubules to water.
c. decreased water reabsorption.
d. increased urine volume.
e. all of the above

_____ 14. Increased aldosterone causes
a. decreased reabsorption of sodium.
b. decreased secretion of potassium.
c. decreased reabsorption of chloride.
d. increased permeability of the distal convoluted tubule to water.
e. decreased volume of urine.

_____ 15. Juxtaglomerular cells secrete
a. ADH.
b. oxytocin.
c. renin.
d. aldosterone.
e. atrial natriuretic hormone.

_____ 16. A lack of ADH results in diabetes insipidus. A person with this condition produces a
a. large volume of concentrated urine.
b. large volume of dilute urine.
c. small volume of concentrated urine.
d. small volume of dilute urine.

_____ 17. Extracellular fluid
a. tends to be higher in sodium and chloride than intracellular fluid.
b. includes interstitial fluid, plasma, and lymph.
c. has a fairly consistent composition throughout the body.
d. all of the above

_____ 18. Which of the following results in increased blood potassium levels?
a. decrease in aldosterone secretion
b. increase in atrial natriuretic hormone secretion
c. increase in renin secretion
d. decrease in ADH secretion

_____19. The sensation of thirst increases when
a. the mouth is dry.
b. the concentration of the blood increases.
c. blood pressure decreases.
d. all of the above

_____20. An increase in blood carbon dioxide levels is followed by a (an) _____ in hydrogen ions, and a (an) _____ in blood pH.
a. decrease, decrease
b. decrease, increase
c. increase, decrease
d. increase, increase

FINAL CHALLENGES

Use a separate sheet of paper to complete this section.

1. What effect would a condition that obstructed blood flow to the kidneys have on renin and aldosterone levels? Explain.

2. Given the following information: alcohol inhibits ADH secretion and caffeine causes increased blood pressure in the glomerular capillaries. Explain the effects on urine production if Mr. I. P. Daily drinks several alcoholic beverages and then tries to sober up by drinking several cups of coffee.

3. Which of the following symptoms are consistent with a diagnosis of excessive aldosterone secretion: polydipsia (excessive drinking), polyuria (excessive urine production), and blood pressure 15% higher than normal? Explain.

4. To be effective, a diuretic (drug that increases water loss through the kidneys) should not only increase water loss, but also increase sodium loss. Why?

5. John Uptight has a gastric ulcer. One day he consumes 10 packages of antacid tablets which are mainly sodium bicarbonate (an alkaline substance). What effect would their consumption have on blood pH, urine pH, and respiration rate?

ANSWERS TO CHAPTER 18

Kidneys
- A. 1. Hilum; 2. Renal sinus; 3. Renal pelvis;
  4. Calyces; 5. Ureter
- B. 1. Cortex; 2. Medulla; 3. Renal pyramids
- C. 1. Cortex; 2. Medulla; 3. Connective tissue
  capsule; 4. Renal pyramid; 5. Ureter; 6. Calyx;
  7. Renal pelvis

The Nephron
- A. 1. Filtrate; 2. Proximal convoluted tubule;
  3. Ascending limb; 4. Distal convoluted tubule;
  5. Collecting duct; 6. Urine
- B. 1. Afferent arteriole; 2. Bowman's capsule;
  3. Glomerulus; 4. Proximal convoluted tubule;
  5. Loop of Henle; 6. Peritubular capillaries;
  7. Collecting duct; 8. Distal convoluted tubule;
  9. Efferent arteriole

Arteries and Veins
1. Renal arteries; 2. Afferent arteriole; 3. Efferent
arteriole; 4. Peritubular capillaries

Ureters, Urinary Bladder, and Urethra
1. Ureters; 2. Urinary bladder; 3. Urethra; 4. Smooth
muscle; 5. External urinary sphincter

Urine Production
- A. 1. Ammonia; 2. Urea
- B. 1. Filtration; 2. Tubular reabsorption; 3. Tubular
  secretion
- C. 1. Filtration membrane; 2. Filtrate; 3. Blood cells;
  4. Water; 5. Filtration pressure; 6. Increases;
  7. Increases; 8. Decreases
- D. 1. Reabsorbed; 2. Osmosis; 3. Peritubular space;
  4. Renal veins; 5. Urine; 6. Waste products;
  7. Proteins and glucose; 8. Secreted

Regulation of Urine Concentration and Volume
- A. 1. Decreases; 2. Increases; 3. Increases;
  4. Increases; 5. Decrease
- B. 1. Increases; 2. Decreases; 3. Increases;
  4. Increases

- C. 1. Increases; 2. Increase; 3. Increases;
  4. Increases; 5. Decreases; 6. Increases;
  7. Increases
- D. 1. Increases; 2. Increases; 3. Decreases;
  4. Increase; 5. Increases; 6. Increase

Urine Movement
1. Spinal cord reflex; 2. Higher brain centers

Body Fluid Compartments
1. Intracellular fluid compartment; 2. Extracellular
fluid compartment; 3. Intracellular fluid
compartment

Regulation of Extracellular Fluid Composition
- A. 1. Increases; 2. Increases; 3. Increases;
  4. Increases
- B. 1. Increased; 2. Increased; 3. Decreased;
  4. Increased; 5. Decreased; 6. Decreased;
  7. Decreased; 8. Increased; 9. Increased;
  10. Decreased; 11. Decreased
- C. 1. Sodium; 2. Kidneys; 3. Potassium;
  4. Aldosterone; 5. Increase; 6. Increases

Regulation of Acid-Base Balance
- A. 1. Protein and phosphate buffer systems;
  2. Bicarbonate buffer system
- B. 1. Rapidly; 2. Carbon dioxide; 3. Decreases;
  4. Increases; 5. Increases; 6. Decreases;
  7. Increase; 8. Slowly; 9. Hydrogen ions;
  10. Increases; 11. Increases

Understanding Acidosis and Alkalosis
1. Acidosis; 2. Central nervous system;
3. Respiratory ; 4. Metabolic; 5. Alkalosis;
6. Hyperexcitability; 7. Respiratory; 8. Metabolic;
9. Kidneys; 10. Respiratory

Disorders of the Urinary System
- A. 1. Acute glomerular nephritis; 2. Chronic
  glomerular nephritis; 3. Cystitis; 4. Kidney stones
- B. 1. Acute renal failure; 2. Chronic renal failure;
  3. Dialysis; 4. Kidney transplant

1. Glomerulus, Bowman's capsule, proximal
   convoluted tubule, loop of Henle, distal convoluted
   tubule
2. Filtration, reabsorption, and secretion
3. Aldosterone: increased sodium ion reabsorption,
   resulting in decreased urine concentration and
   volume; ADH: decreased urine volume; atrial
   natriuretic factor: larger volume of urine with
   greater salt content

4. Stretch of bladder, reflex initiated, bladder contracts,
   and urinary sphincters relax
5. Increased concentration of body fluid, decreased
   blood pressure, and dry mouth
6. ADH, aldosterone, and renin: decreased volume of
   more concentrated urine; increased blood volume
   and blood pressure

7. Bicarbonate buffer system, phosphate buffer system, and protein buffer system

8. Respiration rate above normal: blood pH increases; respiration rate below normal: blood pH decreases

MASTERY LEARNING ACTIVITY

1. E. The glomerulus, Bowman's capsule, and proximal and distal convoluted tubules are found in the renal cortex, whereas the collecting duct and loop of Henle enter the medulla. The renal pelvis is an enlarged urinary channel that between the calyces and the ureter. The Hilum is the indentation on the kidney where the renal arteries, veins, nerves, and ureter attach to the kidney.

2. A. The glomerulus is a tuft of coiled capillaries. Bowman's capsule, the loop of Henle and collecting duct are tubules that contain fluid removed from the blood in the glomerulus by filtration.

3. C. The functional unit of the kidney is the nephron.

4. B. The tipe of each renal pyramid extends into the medulla, where it is surrounded by a calyx. The glomerulus is surrounded by Bowman's capsule, and the loop of Henle is a tubular part of the nephron.

5. B. Filtrate passes into Bowman's capsule, and sequentially through the proximal convoluted tubule, loop of Henle, distal convoluted tubule, and collecting duct.

6. D. Filtration, reabsorption, and secretion are the three processes in urine formation.

7. A. The afferent arteriole passes into the glomerulus, the efferent arteriole passes out of the glomerulus, and extends to the peritubular capillaries.

8. D. The urinary bladder is located in the pelvic cavity. The walls of the urinary bladder are composed of smooth muscle. The urinary bladder is connected to the kidneys by ureters, and to the outside of the body by the urethra.

9. D. Constriction of afferent arterioles reduces blood pressure in the glomerulus, and also reduces filtration pressure. Cardiovascular shock results from a drop in blood pressure, which also reduces filtration pressure. Elevated blood pressure increases filtration pressure.

10. E. All of the substances listed are reabsorbed by the nephron.

11. E. Usually about 99% of the filtrate is reabsorbed, and 1% of the filtrate becomes urine.

12. A. Potassium ions, hydrogen ions, and ammonia are all secreted into the filtrate. Sodium ions are reabsorbed by the nephron.

13. B. ADH increases the permeability of the distal convoluted tubules and collecting ducts to water. This increases water reabsorption and blood volume, and decreases urine volume.

14. E. As aldosterone secretion increases, potassium secretion and sodium and chloride reabsorption increase. As sodium and chloride move from the urine back into the blood water follows by osmosis. This increases the blood volume and decreases urine volume.

15. C. Juxtaglomerular cells are involved in the secretion of renin. ADH and oxytocin are secreted by the posterior pituitary, aldosterone is secreted by the adrenal cortex, and atrial natriuretic hormone is secreted by the atria of the heart.

16. B. With a lack of ADH, the permeability of the collecting tubules decreases, there is less water reabsorbed, producing a larger volume of less concentrated urine.

17. D. Extracellular fluid tends to be higher in sodium, chloride, and bicarbonate ions than intracellular fluid. Extracellular fluid includes interstitial fluid, blood plasma, and lymph, and has a fairly constant composition throughout the body.

18. A. Aldosterone causes increased potassium secretion into the filtrate. A decrease in aldosterone would mean less potassium is secreted, leaving more potassium in the blood. Increased renin increases aldosterone levels, which decreases blood potassium levels. Atrial natriuretic hormone and ADH do not substantially affect blood potassium levels.

19. D. All of these factors trigger the sensation of thirst.

20. C. When blood carbon dioxide levels increase, hydrogen ions also increase. The increase of hydrogen ions lowers the pH, i.e. increases the acidity of the blood.

1. Reduced blood pressure in the afferent arterioles increases renin secretion by the juxtaglomerular apparatus. Renin causes increased production of Angiotensin I, which is converted to angiotensin II. Angiotensin II increases aldosterone secretion, which decreases urine production causing increased blood volume and blood pressure.

2. Urine production increases for several reasons. First, the increased fluid intake increases blood volume, blood pressure, and filtration pressure. Second, caffeine increases filtration pressure, producing a greater amount of filtrate and urine. Third, inhibition of ADH secretion by alcohol decreases the permeability of the collecting ducts to water, so a larger amount of filtrate passes through as urine.

3. Excess aldosterone leads to increased sodium and water reabsorption. Because of the water increase, blood volume increases, causing a rise in blood pressure. Eventually, however, the increase in blood volume is opposed by mechanisms that regulate blood volume/pressure. The increased sodium in the extracellular fluid stimulates the thirst center, resulting in excessive drinking followed by excessive urine production.

4. If only water were lost, blood concentration would increase and stimulate ADH secretion. The ADH would increase the permeability of the collecting ducts to water, increasing water reabsorption and reducing the effectiveness of the diuretic. If sodium is also lost, blood osmolality does not increase, and the ADH response does not occur.

5. The overdose of alkaline antacid tablets raises blood pH, producing metabolic alkalosis. By secreting fewer hydrogen ions, a more alkaline urine is formed, and hydrogen ion concentration in the blood increases, compensating for the alkalosis. Respiration rate also decreases to compensate. Reduced respiration increases plasma CO_2 levels, which reduces the pH through the production of carbonic acid.

The Reproductive System

FOCUS: Reproductive organs in males and females produce sex cells. The reproductive organs sustain the sex cells, transport them to the site where fertilization can occur, and, in the female, nurture the developing offspring both before and, for a time, after birth. Reproductive organs also produce hormones that play important roles in the development and maintenance of the reproductive system. These hormones help determine sexual characteristics, influence sexual behavior, and play a major role in regulating the physiology of the reproductive system.

CONTENT LEARNING ACTIVITY

Formation of Sex Cells

"The formation of sex cells takes place by meiosis."

Match these terms or numbers with the correct statement or definition:

Fertilization	2
Oocyte	4
Polar body	23
Sperm cell	46

_____ 1. The number of cell divisions that occur during meiosis.

_____ 2. The number of chromosomes in human cells before meiosis begins.

_____ 3. The number of chromosomes produced by meiosis in the sex cells of humans.

_____ 4. In females, the developing sex cell that receives most of the cytoplasm.

_____ 5. Occurs when a male and female sex cell unite.

Testes and Scrotum

66_The testes are the male's primary reproductive organs._**99**

Match these terms with the
correct statement or definition:

Abdominal muscle extensions   Lobules
Efferent ductules   Scrotum
Inguinal canal   Seminiferous tubules
Inguinal hernia   Smooth muscle of scrotum
Interstitial cells

_____ 1. Sac containing the testes.

_____ 2. Two structures that regulate the temperature of the testes.

_____

_____ 3. Subdivisions of the testes.

_____ 4. Site of sperm cell development.

_____ 5. Responsible for testosterone production.

_____ 6. Tubes that exit the testes.

_____ 7. Passageway used by the testis to pass through the abdominal wall.

_____ 8. Protrusion of the small intestine through the inguinal canal.

☞   If the testes become too warm or too cold, normal sperm cell development does not occur.

Spermatogenesis

66_Spermatogenesis is the formation of sperm cells._**99**

Match these terms with the
correct statement or definition:

Primary spermatocytes   Sperm cell
Secondary spermatocytes   Spermatid
Sertoli cells   Spermatogonia

_____ 1. Large cells that nourish and support the germ cells.

_____ 2. Most peripheral cells; they divide by mitosis.

_____ 3. Germ cells produced from spermatogonia, which divide into two cells during the first meiotic division.

_____ 4. Formed from primary spermatocytes these cells undergo a second meiotic division.

_____ 5. Produced from secondary spermatocytes; each has 23 chromosomes.

_____ 6. Develop from a spermatid by forming a head, midpiece, and flagellum.

Ducts

"*Sperm cells leave the testes and pass through a series of ducts to reach the exterior of the body.*"

Match these terms with the correct statement or definition:

Ductus deferens   Spermatic cord
Ejaculatory duct   Urethra
Epididymis

_____ 1. Receives the efferent ductules from the testis; a comma-shaped structure on the outside of the testis.

_____ 2. Site at which sperm cells become capable of fertilization.

_____ 3. Passes through the inguinal canal.

_____ 4. The ductus deferens and blood vessels and nerves that supply the testis.

_____ 5. Formed by the ductus deferens and a duct from the seminal vesicle; empties into the urethra.

_____ 6. Extends to the distal end of the penis.

A vasectomy is a surgical procedure for producing sterility in males. An incision through the scrotum exposes the ductus deferens within the spermatic cord. The ductus deferens are cut and their ends are tied off, preventing sperm cells from the testes from exiting the body.

Penis

"*The penis is the male organ of copulation and it transfers sperm cells from the male to the female.*"

Match these terms with the correct statement or definition:

Erection   Corpus spongiosum
Circumcision   Glans penis
Corpora cavernosa   Prepuce

_____ 1. Engorgement of penile erectile tissue with blood.

_____ 2. Paired columns of erectile tissue in the penis.

_____ 3. Single column of erectile tissue in the penis; the urethra passes through it.

_____ 4. Expanded distal end of the penis.

_____ 5. Skin that covers the glans penis; foreskin.

_____ 6. Surgical removal of the prepuce.

Glands and Secretions

" *Several glands secrete substances into the ducts of the reproductive system.* **"**

A. Match these terms with the correct statement or definition:

Bulbourethral glands   Semen
Prostate   Seminal vesicles

_____

_____

_____

_____

_____

_____

_____

1. Mixture of sperm cells and secretions from male reproductive glands.

2. Two sac-shaped glands adjacent to the ductus deferens where it approaches the prostate gland.

3. Gland the size and shape of a walnut that surrounds the urethra and the two ejaculatory ducts.

4. Small pair of glands located near the base of the penis.

5. Glands producing a mucous secretion that neutralizes the acidic urethra; secreted up to several minutes before ejaculation.

6. Glands producing thick, mucoid secretions containing nutrients that nourish the sperm cells.

7. Gland producing thin, milky secretions that neutralizes acidic secretions of the seminal vesicles and the vagina.

B. Match these terms with the correct parts labeled in Figure 19-1:

Bulbourethral gland
Ductus deferens
Ejaculatory duct
Epididymis
Penis
Prostate
Scrotum
Seminal vesicle
Testis
Urethra

1. _____

2. _____

3. _____

4. _____

5. _____

6. _____

7. _____

8. _____

9. _____

10. _____

Physiology of Male Reproduction

66*Hormones are primarily responsible for the development and maintenance of reproductive structures.*99

Match these terms with the correct statement or definition:

Anabolic steroids   LH
FSH   Testosterone
GnRH

_____

1. Hormone released from the hypothalamus; stimulates the anterior pituitary to secrete two hormones.

_____

2. Released from the anterior pituitary; stimulates interstitial cells to secrete testosterone.

_____

3. Released from the anterior pituitary; influences Sertoli cells and promotes spermatogenesis.

_____

4. During puberty promotes the development of secondary sexual characteristics; after puberty maintains adult structures; necessary for spermatogenesis.

_____

5. Hormone important in maintaining the male sexual drive.

_____

6. Sometime used by athletes in an attempt to increase muscle mass; has many undesirable side effects.

Male Sexual Behavior and the Male Sex Act

66*Neural mechanisms are primarily involved in controlling the sexual act.*99

Match these terms with the correct statement or definition:

Ejaculation   Orgasm
Emission   Resolution
Erection

_____

1. Pleasurable sensation associated with ejaculation.

_____

2. Occurs when the arteries that supply blood to the erectile tissue of the penis dilate.

_____

3. Release of the secretions of the seminal vesicles and prostate gland into the urethra.

_____

4. Semen is forced out of the urethra by rhythmic contractions of skeletal muscles at the base of the penis.

_____

5. An overall feeling of satisfaction; the male is unable to achieve erection and a second ejaculation.

Ovaries

"The ovaries are the female's primary reproductive organs."

A. Match these terms with the correct statement or definition:

Mature follicle   Primary oocyte
Oogonia   Secondary follicle
Ovarian follicle   Secondary oocyte
Primary follicle

_____ 1. Cells that give rise to oocytes; divide by mitosis.

_____ 2. Oocyte that has started the first meiotic division.

_____ 3. Oocyte that has started the second meiotic division.

_____ 4. General term for an oocyte and the cells that surround it.

_____ 5. Consists of a layer of cells surrounding the primary oocyte.

_____ 6. Follicle that has developed an antrum.

_____ 7. Enlarged follicle on the surface of the ovary.

☞ The developing follicles secrete estrogen that prepares the uterus to receive the fertilized oocyte.

B. Using the terms provided, complete the following statements:

Corpus luteum   Placenta
Estrogen   Progesterone
Fertilization   Secondary oocyte
Ovulation

Rupture of the mature follicle and release of the oocyte from the ovary is called (1). Near this time the primary oocyte completes the first meiotic division to form the (2), which begins the second meiotic division. The second meiotic division is completed only if (3) occurs. The ruptured follicle becomes the (4), which secretes (5) and (6). If fertilization occurs, the corpus luteum persists and continues to produce hormones that are necessary to maintain the pregnancy. After the first trimester the (7) produces hormones and the corpus luteum of pregnancy degenerates.

1. _____

2. _____

3. _____

4. _____

5. _____

6. _____

7. _____

Uterine Tubes, Uterus, and Vagina

❝*The uterus is the site of development of a new individual.***❞**

A. **M**atch these terms with the correct statement or definition:

Body   Myometrium
Cervical canal   Serous layer
Cervix   Uterine cavity
Endometrium   Uterine tubes
Hymen   Vagina

_____ 1. Extend from the ovaries to the uterus; also called fallopian tubes or oviducts.

_____ 2. Fertilization usually occurs here.

_____ 3. Narrow, inferior part of the uterus.

_____ 4. Space that opens into the vagina.

_____ 5. Middle muscular layer of the uterine wall.

_____ 6. Inner epithelial and connective tissue layer of the uterus.

_____ 7. Receives the penis during intercourse and allows menstrual flow and childbirth.

_____ 8. Mucous membrane covering the opening of the vagina.

B. Match these terms with the correct parts labeled in Figure 19-2:

Body
Cervical canal
Cervix
Endometrium
Myometrium
Ovary
Serous layer
Uterine cavity
Uterine tube
Vagina

1. _____ 5. _____ 8. _____

2. _____ 6. _____ 9. _____

3. _____ 7. _____ 10. _____

4. _____

External Genitalia

"_The external genitalia is also called the vulva._**"**

A. Match these terms with the correct statement or definition:

Clinical perineum
Clitoris
Labia majora
Labia minora

Mons pubis
Vestibular glands
Vestibule

_____ 1. The space into which the vagina and urethra open.

_____ 2. Thin, longitudinal skin folds bordering the vestibule.

_____ 3. Small erectile structure containing a large number of sensory structures.

_____ 4. Glands that maintain the moistness of the vestibule.

_____ 5. Rounded folds of skin lateral to the labia minora.

_____ 6. Mound located over the pubic symphysis.

_____ 7. The region between the vagina and anus; the location where an episiotomy is performed.

B. Match these terms with the correct parts labeled in Figure 19-3:

Clinical perineum
Clitoris
Labia majora
Labia minora
Mons pubis
Prepuce
Urethra
Vagina
Vestibule

1. _____ 4. _____ 7. _____

2. _____ 5. _____ 8. _____

3. _____ 6. _____ 9. _____

Mammary Glands

"*The mammary glands are the organs of milk production located in the breasts.***"**

A. Match these terms with the correct statement or definition:

Alveoli	Lobe
Areola	Oxytocin
Colostrum	Parturition
Estrogen	Progesterone
Lactation	Prolactin

_____ 1. Circular, pigmented area surrounding the nipple.

_____ 2. Glandular compartment of the mammary glands, each of which possesses a single duct that opens on the surface of the nipple.

_____ 3. Secretory sacs that produce milk.

_____ 4. The production of milk.

_____ 5. Produced for the first few days after childbirth; contains little fat and less lactose than milk.

_____ 6. Two hormones that stimulate fat deposition, duct development, and alveoli development in the breasts.

_____

_____ 7. Hormone that stimulates milk production.

_____ 8. Hormone that stimulates milk letdown.

B. Match these terms with with the correct parts labeled in Figure 19-4:

Areola	Lobe
Duct	Nipple
Fat	

1. _____

2. _____

3. _____

4. _____

5. _____

Puberty

Puberty in females is marked by the first episode of menstrual bleeding, which is called menarche.

Using the terms provided, complete the following statements:

Cyclic   GnRH
Estrogen and progesterone   High
FSH and LH   Low

1. _____

2. _____

3. _____

4. _____

The changes associated with puberty in the female are primarily the result of elevated levels of _(1)_ secreted by the ovaries. Before puberty, the rate of secretion of _(2)_ from the hypothalamus, and _(3)_ from the anterior pituitary are very _(4)_. After puberty the rate of secretion of GnRH, FSH, and LH increases, and is responsible for the pattern of estrogen and progesterone secretion of the adult.

Menstrual Cycle

The term menstrual cycle refers to the series of changes that occur in sexually mature, nonpregnant women that results in menses.

A. Match these terms with the correct statement or definition:

Menses   Premenstrual syndrome
Menopause   Proliferative phase
Ovulation   Secretory phase

_____ 1. Phase in which the endometrium of the uterus is sloughed; day 1 to days 4 or 5 of the cycle.

_____ 2. Phase in which the endometrium begins to thicken and form tubelike glands; days 4 or 5 to day 14 of the cycle.

_____ 3. Release of the oocyte from the ovary; day 14 of the cycle.

_____ 4. Phase in which the endometrium reaches its maximum thickness and glands secrete a small amount of fluid; days 14 to 28 of the cycle.

_____ 5. Phase in which a mature follicle is produced.

_____ 6. Phase in which the corpus luteum is formed.

_____ 7. Severe changes in mood just before menses.

_____ 8. Cessation of menstrual cycles.

☞ Prior to menopause some women experience "hot flashes," irritability, fatigue, anxiety, and occasionally severe emotional disturbances. Many of these symptoms can be effectively treated with estrogen.

B. Match these terms with the correct statement or definition:

Estrogen   LH
FSH   Progesterone
GnRH

_____   1. Hormone that stimulates follicle development.

_____   2. Increased secretion of this hormone from developing follicles causes the endometrium to thicken during the proliferative phase.

_____   3. Stimulates the secretion of FSH and LH from the anterior pituitary.

_____   4. Two hormones that stimulate the secretion of estrogen from follicles.

_____

_____   5. Hormone that stimulates ovulation.

_____   6. Produced by the corpus luteum; primarily responsible for the secretory phase of the uterus.

_____   7. Decline in this hormone causes the endometrium to be sloughed and results in menses.

C. Match these terms with the correct parts labeled in Figure 19-5 on the next page:

Corpus luteum   Menses
Degenerated corpus luteum   Progesterone
Estrogen   Proliferative phase
GnRH   Secondary follicle
Mature follicle   Secretory phase

1. _____   5. _____   9. _____

2. _____   6. _____   10. _____

3. _____   7. _____   11. _____

4. _____   8. _____

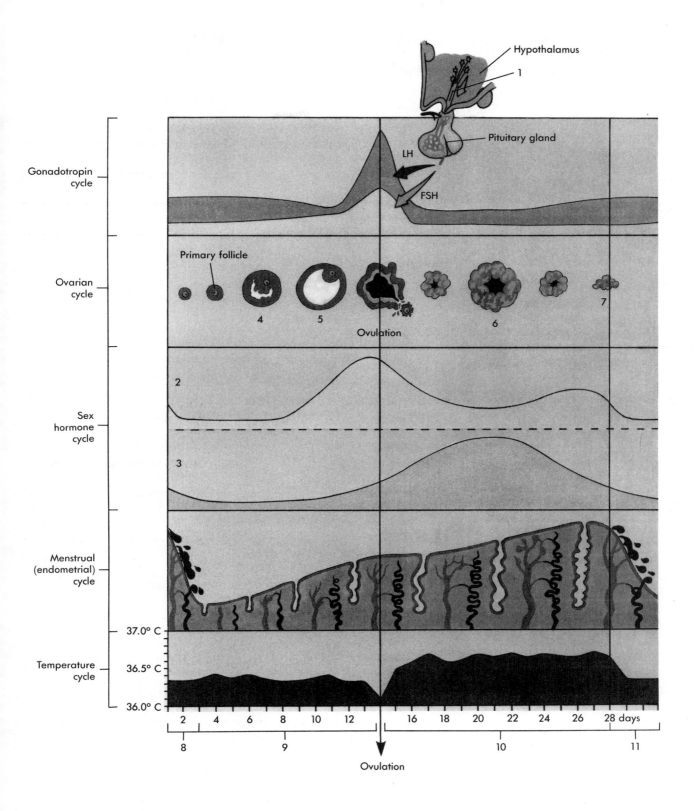

Gonadotropin cycle

Hypothalamus

1

Pituitary gland

LH

FSH

Ovarian cycle

Primary follicle

4

5

Ovulation

6

7

Sex hormone cycle

2

3

Menstrual (endometrial) cycle

Temperature cycle

37.0° C

36.5° C

36.0° C

2   4   6   8   10   12   16   18   20   22   24   26   28 days

8   9   10   11

Ovulation

Female Sexual Behavior and the Female Sex Act

❝_Sexual drive in females, like sexual drive in males, is dependent upon hormones._**❞**

Using the terms provided, complete the following statements:

Clitoris   Resolution
Fertilization   Vagina
Orgasm

Testosterone-like hormones and possibly estrogens affect sexual behavior. During sexual excitement, erectile tissue in the _(1)_ and around the vaginal opening become engorged with blood. Secretions from the _(2)_ provide lubrication for the movement of the penis. Tactile stimulation during intercourse, as well as psychological stimuli normally triggers a(n) _(3)_, the female climax. After the sexual act, there is a period of _(4)_, characterized by an overall sense of satisfaction and relaxation. Although orgasm is a pleasurable component of sexual intercourse, it is not required for _(5)_ to occur.

1. _____
2. _____
3. _____
4. _____
5. _____

Reproductive Disorders

❝_Knowledge of reproductive orders is clinically important._**❞**

A. Match these terms with the correct statement or definition:

AIDS   Pelvic inflammatory disease
Genital herpes   Syphilis
Gonorrhea   Trichomonas
Nongonococcal urethritis

_____ 1. Protozoan infection of the vagina of females and the urethra of males; more common in females.

_____ 2. Bacterial infection that can cause sterility; males typically discharge pus from the urethra.

_____ 3. Any infection of the urethra not caused by the gonorrhea bacteria.

_____ 4. Infection by herpes simplex type 2 virus that produces blister-like areas of inflammation.

_____ 5. Bacterial infection that has three stages: stage one produces small sores at the site of infection; stage two causes skin rashes and fever; stage three damages the nervous system.

_____ 6. Viral infection that depresses the immune system.

_____ 7. General term for bacterial infections of the pelvic organs.

Understanding the Prevention of Pregnancy

"*In our society, methods that prevent pregnancy are very important.***"**

A. **M**atch these terms with the correct statement or definition:

Abortion	Diaphragm
Coitus interruptus	Douche
Condom	Rhythm method
Contraception	Spermicidal agent

_____ 1. General term for the prevention of fertilization.

_____ 2. The removal of the embryo.

_____ 3. Removal of the penis just before ejaculation.

_____ 4. Abstaining from sexual intercourse near the time of ovulation.

_____ 5. A sheath placed over the penis or in the vagina.

_____ 6. A plastic or rubber dome placed over the cervix.

_____ 7. Chemicals used to kill sperm cells; often in a sponge placed over the cervix.

_____ 8. A device used to rinse the vagina with a chemical that kills sperm cells; not very effective.

B. **M**atch these terms with the correct statement or definition:

Implant	RU486
IUD	Tubal ligation
Oral contraceptive	Vasectomy

_____ 1. Synthetic estrogen and progesterone taken orally to suppress fertility.

_____ 2. Progesterone-like substance placed beneath the skin.

_____ 3. Surgical procedure that cuts and ties off the ductus deferens.

_____ 4. Surgical procedure that cuts and ties off the uterine tubes.

_____ 5. Drug that blocks the action of progesterone and induces abortion.

_____ 6. Device placed in the uterus that prevents implantation.

1. List in the order of their formation the cells that are formed during spermatogenesis.

2. Starting at the site of sperm cell production, name in order the ducts sperm cells would pass through to reach the exterior of the body.

3. Name the three glands in the male reproductive system and describe their secretions.

4. State the functions of GnRH, FSH, and LH in males.

5. List the functions of testosterone in the males.

6. List in the order of their formation the oocytes and follicles of the ovary. Name the structure that develops from the follicle after ovulation.

7. List and the give the functions of four hormones that affect the breasts, mammary glands, and lactation.

8. Name the three phases of the menstrual cycle.

9. State the functions of GnRH, FSH, and LH in females.

10. List the functions of estrogen and progesterone in females.

MASTERY LEARNING ACTIVITY

Place the letter corresponding to the correct answer in the space provided.

_____ 1. Which of the following is correctly
matched?
a. sperm cell - 46 chromosomes
b. oocyte - receives most of the
cytoplasm
c. polar body - sun bathing Eskimo
d. fertilization - union of sperm cell and
polar body

_____ 2. If an adult male jumped into a
swimming pool of cold water, which of
the following would be expected to
happen?
a. extensions of the abdominal muscles
relax
b. smooth muscle in the scrotum
contracts
c. the skin of the scrotum becomes loose
and thin
d. the testes descend away from the
body

_____ 3. Which of the following is correctly
matched with its function?
a. interstitial cells (cells of Leydig) -
testosterone production
b. Sertoli cells - nourish developing
sperm cells
c. seminiferous tubules - site of
spermatogenesis
d. all of the above

_____ 4. Given the following structures:
1. ductus deferens
2. efferent ductule
3. epididymis
4. ejaculatory duct

Choose the arrangement that lists the
structures in the order sperm cells pass
through them.
a. 2, 3, 1, 4
b. 2, 3, 4, 1
c. 3, 2, 1, 4
d. 3, 2, 4, 1

_____ 5. Concerning the penis,
a. The urethra passes through the
corpora cavernosa.
b. the glans penis is formed by the
corpora cavernosa.
c. the prepuce covers the glans penis.
d. all of the above

_____ 6. Given the following glands:
1. prostate gland
2. bulbourethral glands
3. seminal vesicles

Choose the arrangement in which the
glands would contribute their secretions
during the formation of semen.
a. 1, 2, 3
b. 2, 1, 3
c. 2, 3, 1
d. 3, 1, 2
e. 3, 2, 1

_____ 7. Which of the following glands is correctly matched with the function of the gland's secretion?
a. bulbourethral glands - neutralizes acidic contents of the urethra
b. seminal vesicles - contains nutrients that nourish the sperm cells
c. prostate gland - neutralizes the acidic secretions of the seminal vesicles and vagina
d. all of the above

_____ 8. LH in the male
a. stimulates GnRH secretion.
b. stimulates Sertoli cells to divide.
c. is higher before puberty than after puberty.
d. stimulates testosterone production.

_____ 9. Which of the following is correctly matched?
a. erection - venous sinuses fill with blood
b. emission - semen accumulates in the ductus deferens
c. ejaculation - contraction of skeletal muscles in the ejaculatory duct
d. all of the above

_____ 10. The primary follicle
a. contains a primary oocyte.
b. contains an antrum.
c. forms a lump on the surface of the ovary.
d. all of the above

_____ 11. The corpus luteum
a. is formed from a primary follicle.
b. produces large amounts of testosterone.
c. degenerates in a few days if fertilization occurs.
d. functions until the placenta produces progesterone.

_____ 12. Given the following structures:
1. cervical canal
2. peritoneal cavity
3. uterine cavity
4. uterine tube

Assume a couple has just consummated the sex act and sperm cells of have been deposited in the vagina. Trace the pathway of the sperm cells through the female's reproductive tract to the ovary.
a. 1, 3, 2, 4
b. 1, 3, 4, 2
c. 3, 1, 2, 4
d. 3, 1, 4, 2
e. 4, 2, 1, 3

_____ 13. Given the following structures:
1. vaginal opening
2. clitoris
3. urethral opening
4. anus

Choose the arrangement that lists the structures in their proper order from the anterior to the posterior aspect.
a. 2, 3, 1, 4
b. 2, 4, 3, 1
c. 3, 1, 2, 4
d. 3, 1, 4, 2
e. 4, 2, 3, 1

_____ 14. Concerning the breasts,
a. the alveoli are the site of milk production.
b. ducts from the mammary glands open on the areola.
b. the female breast enlarges in response to oxytocin.
d. prolactin stimulates milk letdown.

_____ 15. The layer of the uterus that undergoes the greatest change during the menstrual cycle is the
a. endometrium
b. hymen
c. myometrium
d. serous layer

_____16. The major secretory product of the mature follicle is
   a. estrogen.
   b. progesterone.
   c. LH.
   d. FSH.
   e. GnRH.

_____17. Which of the following processes or phases in the menstrual cycle occur at the same time?
   a. proliferative phase of the uterus and increased estrogen production by the ovary
   b. maximal LH secretion and menstruation

   c. degeneration of the corpus luteum and an increase in ovarian progesterone production
   d. menstruation and an increase in ovarian progesterone production
   e. ovulation and menstruation

_____18. Menopause
   a. happens whenever a woman pauses to think about a man.
   b. occurs when a woman stops a man from making a pass.
   c. develops when follicles become less responsive to FSH and LH.
   d. results from high estrogen levels in 40 - 50 year old women.

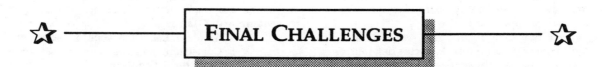

☆ ——————— **FINAL CHALLENGES** ——————— ☆

Use a separate sheet of paper to complete this section.

1. What would happen to testosterone production in the testes in response to an injection of a large amount of testosterone in an adult male? Explain.

2. Suppose a 9 year-old boy had an interstitial cell tumor that resulted in very high levels of testosterone production. Describe the effects this would have on his development.

3. Birth control pills that consist of estrogen or progesterone are only taken for 21 days. The woman stops taking the birth control pill or takes a placebo pill for 7 days. Then she resumes taking the birth control pill. Why does she do this?

4. A woman who was taking birth control pills that consisted of only progesterone experienced the hot flash symptoms of menopause. Explain.

5. Sexually transmitted diseases such as gonorrhea can sometimes cause peritonitis in females. In males, however, sexually transmitted diseases do not cause peritonitis. Explain.

ANSWERS TO CHAPTER 19

CONTENT LEARNING ACTIVITY

Meiosis
1. 2; 2. 46; 3. 23; 4. Oocyte; 5. Fertilization

Testes and Scrotum
1. Scrotum; 2. Abdominal muscle extensions and smooth muscle of scrotum; 3. Lobules;
4. Seminiferous tubules; 5. Interstitial cells;
6. Efferent ductules; 7. Inguinal canal; 8. Inguinal hernia

Spermatogenesis
1. Sertoli cells; 2. Spermatogonia; 3. Primary spermatocytes; 4. Secondary spermatocytes;
5. Spermatid; 6. Sperm cell

Ducts
1. Epididymis; 2. Epididymis; 3. Ductus deferens;
4. Spermatic cord; 5. Ejaculatory duct; 6. Urethra

Penis
1. Erection; 2. Corpora cavernosa; 3. Corpus spongiosum; 4. Glans penis; 5. Prepuce;
6. Circumcision

Glands and Secretions
A. 1. Semen; 2. Seminal vesicles; 3. Prostate;
4. Bulbourethral glands; 5. Bulbourethral glands;
6. Seminal vesicles; 7. Prostate
B. 1. Urethra; 2. Penis; 3. Scrotum; 4. Testis;
5. Epididymis; 6 Ductus deferens;
7. Bulbourethral gland; 8. Prostate; 9. Ejaculatory duct; 10. Seminal vesicle

Physiology of Male Reproduction
1. GnRH; 2. LH; 3. FSH; 4. Testosterone;
5. Testosterone; 6. Anabolic steroids

Male Sexual Behavior and the Male Sex Act
1. Orgasm; 2. Erection; 3. Emission; 4. Ejaculation;
5. Resolution

Ovaries
A. 1. Oogonia; 2. Primary oocyte; 3. Secondary oocyte; 4. Ovarian follicle; 5. Primary follicle;
6. Secondary follicle; 7. Mature follicle
B. 1. Ovulation; 2. Secondary oocyte; 3. Fertilization;
4. Corpus luteum; 5. Estrogen; 6. Progesterone;
7. Placenta

Uterine Tubes, Uterus, and Vagina
A. 1. Uterine tubes; 2. Uterine tubes; 3. Cervix;
4. Cervical canal; 5. Myometrium;
6. Endometrium; 7. Vagina; 8. Hymen

B. 1. Body of uterus; 2. Cervix; 3. Vagina; 4. Cervical canal; 5. Serous layer; 6. Myometrium;
7. Endometrium; 8. Uterine tube; 9. Ovary;
10. Uterine cavity

External Genitalia
A. 1. Vestibule; 2. Labia minora; 3. Clitoris;
4. Vestibular glands; 5. Labia majora; 6. Mons pubis; 7. Clinical perineum
B. 1. Prepuce; 2. Labia minora; 3. Vagina;
4. Vestibule; 5. Clinical perineum; 6. Labia majora; 7. Urethra; 8. Clitoris; 9. Mons pubis

Mammary Glands
A. 1. Areola; 2. Lobe; 3. Alveoli; 4. Lactation;
5. Colostrum; 6. Estrogen and progesterone;
7. Prolactin; 8. Oxytocin
B. 1. Duct; 2. Nipple; 3. Areola; 4. Fat; 5. Lobe

Puberty
1. Estrogen and progesterone; 2. GnRH; 3. FSH and LH; 4. Low

Menstrual Cycle
A. 1. Menses; 2. Proliferative phase; 3. Ovulation;
4. Secretory phase; 5. Proliferative phase;
6. Secretory phase; 7. Premenstrual syndrome;
8. Menopause;
B. 1. FSH; 2. Estrogen; 3. GnRH; 4. FSH and LH;
5. LH; 6. Progesterone; 7. Progesterone
C. 1. GnRH; 2. Estrogen; 3. Progesterone;
4. Secondary follicle; 5. Mature follicle; 6. Corpus luteum; 7. Degenerated corpus luteum;
8. Menses; 9. Proliferative phase; 10. Secretory phase; 11. Menses

Female Sexual Behavior and the Female Sex Act
1. Clitoris; 2. Vagina; 3. Orgasm; 4. Resolution;
5. Fertilization

Reproductive Disorders
1. Trichomonas; 2. Gonorrhea; 3. Nongonococcal urethritis; 4. Genital herpes; 5. Syphilis; 6. AIDS;
7. Pelvic inflammatory disease

Understanding the Prevention of Pregnancy
A. 1. Contraception; 2. Abortion; 3. Coitus interruptus; 4. Rhythm method; 5. Condom;
6. Diaphragm; 7. Spermicidal agent; 8. Douche
B. 1. Oral contraceptive; 2. Implant; 3. Vasectomy;
4. Tubal ligation; 5. RU486; 6. IUD

1. Spermatogonia, primary spermatocytes, secondary spermatocytes, spermatids, sperm cells
2. Seminiferous tubule, efferent ductule, ductus deferens, ejaculatory duct, urethra
3. Seminal vesicles: thick mucoid secretion that nourishes sperm cells; Prostate: thin, milky secretion that neutralizes seminal vesicle secretions and the vagina; Bulbourethral glands: mucous secretion that neutralizes the acidic urethra
4. In males GnRH stimulates the release of FSH and LH. FSH stimulates spermatogenesis, and LH stimulates testosterone secretion
5. Testosterone causes the male reproductive organs and ducts to develop; it is necessary for spermatogenesis and it is responsible for secondary sexual characteristics; increases sex drive
6. Oogonia, primary oocyte, secondary oocyte. Primary follicle, secondary follicle, mature follicle, which becomes the corpus luteum after ovulation
7. Estrogen and progesterone promote fat deposition and duct development; prolactin stimulates milk production; and oxytocin stimulates milk letdown
8. Menses, proliferative phase, and secretory phase
9. In females GnRH stimulates the release of FSH and LH. FSH stimulates follicle development, and LH stimulates ovulation
10. Both are involved with the changes that occur at puberty; estrogen promotes the proliferative phase of the menstrual cycle and progesterone the secretory phase; decreasing progesterone levels induce menses; estrogens may affect sexual behavior

1. B.   The oocyte receives most of the cytoplasm and the polar body receives very little. Fertilization is the union of the sperm cell with a secondary oocyte. Sperm cells have 23 chromosomes as a result of meiosis.

2. B.   The smooth muscle of the scrotum and the extensions of the abdominal muscles contract. The scrotum decreases in size and the testes are pulled closer to the body.

3. D.   The interstitial cells produce testosterone, the Sertoli cells nourish the developing sperm cells, and the seminiferous tubules are the site of spermatogenesis.

4. A.   The sperm cells would pass through the efferent ductule, epididymis, ductus deferens, and ejaculatory duct.

5. C.   The prepuce covers the glans penis, which is an expansion of the corpus spongiosum. The urethra passes through the corpus spongiosum.

6. D.   The seminal vesicles empty into the ejaculatory duct, the prostate into the urethra, and the bulbourethral glands into the urethra.

7. D.   All of the secretions are correctly matched with their functions.

8. D.   LH in the male stimulates testosterone secretion from the interstitial cells. LH levels are lower before puberty than after.

9. A.   Erection occurs when venous sinuses in the penis fill with blood. Emission is the accumulation of semen in the urethra. Ejaculation results from the contraction of skeletal muscles at the base of the penis.

10. A.   The primary follicle contains the primary oocyte. The secondary follicle has an antrum, and the mature follicle forms a lump on the surface of the ovary.

11. D.   The corpus luteum is formed from a mature follicle after it ruptures during ovulation. The corpus luteum produces progesterone and smaller amounts of estrogen. It degenerates in a few days if fertilization does NOT occur. If fertilization occurs, it persists and produces hormones until the end of the first trimester, at which time the placenta takes over the production of hormones and the corpus luteum degenerates.

12. B.   The sperm cells pass through the cervical canal into the uterine cavity. From the uterus they enter the uterine tube and pass into the peritoneal cavity.

13. A.   The order is clitoris, urethral opening, vaginal opening, and anus.

14. A.   The alveoli are the site of milk production. The ducts from the mammary gland lobes open on the nipple. Prolactin stimulates milk production and oxytocin stimulates milk letdown. Estrogen and progesterone cause breast enlargement.

15. A. The endometrium is epithelial tissue that undergoes the greatest change during the menstrual cycle.

16. A. The major secretory product of the mature follicle is estrogen. The corpus luteum secretes progesterone and smaller amounts of estrogen. FSH and LH are secreted by the anterior pituitary, and GnRH is secreted by the hypothalamus.

17. A. Increased estrogen levels produced by the developing follicles stimulate the endometrium to proliferate.

18. C. Menopause results from a decreased sensitivity of the follicles to FSH and LH. Estrogen levels in menopause are low because the follicles no longer develop and secrete estrogen.

☆ FINAL CHALLENGES ☆

1. The injected testosterone, by a negative-feedback mechanism, inhibits the production of GnRH from the hypothalamus. Decreased GnRH results in decreased LH release from the anterior pituitary, which causes a reduction in the production of testosterone by the testes.

2. The testosterone would cause early and pronounced development of his sexual organs. He would also have rapid growth of muscle and bone.

3. Estrogen and/or progesterone levels fall when the birth control pill is not taken, which causes menstruation.

4. Lack of estrogen can result in the hot flash symptoms of menopause. Normally FSH from the anterior pituitary stimulates the ovaries to produce estrogen. However, this is prevented by the progesterone only birth control pill. The progesterone inhibits GnRH secretion from the hypothalamus, and this inhibits FSH and LH.

5. In females the microorganisms can travel from the vagina to the uterus, to the uterine tubes, to the peritoneal cavity. Infection of the peritoneum results in peritonitis. In males the microorganisms can move up the urethra to the bladder or into the ejaculatory duct to the ductus deferens. There is no direct connection to the peritoneal cavity in the male, so peritonitis does not develop.

Development and Heredity

FOCUS: The prenatal period begins with fertilization. The developing organism is considered an embryo from the end of the second week to the end of the eighth week; development of the organs occurs during this period. During the last 7 months weight and size increase, and the developing organism is called a fetus. Parturition, or delivery, occurs in three stages, and is influenced by several factors. Human genetics, the study of inherited human traits, is necessary to understand, predict, and prevent genetic disorders.

CONTENT LEARNING ACTIVITY

Development

66 *Events that occur during the prenatal period have profound effects on the rest of a person's life.* **99**

A. Match these terms with the correct statement or definition:

Blastocyst
Fertilization
Fraternal twins
Identical twins
Inner cell mass
Zygote

_____ 1. The union of a sperm cell with an oocyte.

_____ 2. A single cell formed from the union of a sperm cell with an oocyte.

_____ 3. A hollow ball of cells formed 3 or 4 days after fertilization.

_____ 4. Collection of cells at one end of the blastocyst, some of which give rise to the developing organism.

_____ 5. Two organisms formed when one oocyte is fertilized by one sperm cell, followed by separation of the developing cells.

_____ 6. Two organisms formed when two oocytes are each fertilized by a different sperm cell.

B. Match these terms with the correct statement or definition:

Amniotic cavity   Fetus
Embryo   Primitive streak
Embryonic disk   Yolk sac

_____ 1. Cavity within the inner cell mass that produces blood cells.

_____ 2. Cavity within the inner cell mass that becomes a fluid-filled sac into which the developing organism will grow.

_____ 3. A flat disk of tissue between the yolk sac and amniotic cavity.

_____ 4. A thickened line formed from proliferating ectoderm cells that migrate toward the center of the embryonic disk.

_____ 5. Name for the developing organism from the end of the second week to the end of the eighth week of development.

_____ 6. Name for the developing organism during the last 7 months of the prenatal period.

C. Match these terms with the correct statement or definition:

Ectoderm   Germ layers
Endoderm   Mesoderm

_____ 1. Collectively, ectoderm, endoderm, and mesoderm.

_____ 2. The layer of cells in the embryo that gives rise to the linings of the digestive tract, respiratory tract, and urinary tract.

_____ 3. The layer of cells in the embryo that gives rise to most muscle, most connective tissue, blood vessels and reproductive organs.

_____ 4. The layer of cells in the embryo that give rise to the epidermis and glands of the skin, and the brain, spinal cord, and nerves.

D. Using the terms provided, complete the following statements:

Chorion
Chorionic villi
Human chorionic
    gonadotropin (HCG)
Implantation
Placenta
Umbilical cord
Uterine tube

1. _____

2. _____

3. _____

4. _____

5. _____

6. _____

7. _____

Fertilization normally occurs in the _(1)_. About 7 days after ovulation, the uterus is prepared for _(2)_, the attachment of the blastocyst to the uterine wall. The single layer of cells that form the wall of the blastocyst become the _(3)_. The lining of the uterus and the chorion form the _(4)_, which functions to exchange nutrients, gases, and waste products between the mother and the embryo. The connecting stalk between the embryo and placenta elongates and becomes known as the _(5)_. Fetal blood vessels are inside fingerlike projections called _(6)_ that extend into cavities containing pools of maternal blood within the uterine wall. The chorion secretes _(7)_, which stimulates the corpus luteum in the ovary to produce estrogen and progesterone.

Most pregnancy tests are designed to detect HCG in either urine or blood.

Parturition

❝ *Parturition refers to the process by which the baby is born.* **❞**

A. Match these terms with the correct statement or definition:

First stage of labor
Second stage of labor

Third stage of labor

_____ 1. From onset of regular contractions until the cervix dilates to 10 cm.

_____ 2. From maximal cervical dilation until the baby exits the vagina.

_____ 3. Expulsion of the placenta from the uterus.

B. Using the terms provided, complete the following statements:

Adrenocorticotropic hormone (ACTH)
Adrenal steroids

Contractions
Oxytocin
Prostaglandins

1. _____

2. _____

3. _____

4. _____

5. _____

Many factors that support parturition have been identified. Stress caused by the increased size of the fetus within the uterus results in _(1)_ secretion from the fetal pituitary gland. ACTH stimulates the release of _(2)_ which cause the placenta to increase secretions of estrogen and _(3)_, both of which increase uterine contractility. In addition, stretch of the uterus stimulates the release of _(4)_ from the mother's pituitary gland. Oxytocin also increases uterine _(5)_.

Disorders of Pregnancy

❝ *During pregnancy, disorders may affect the mother and/or the fetus.* **❞**

A. Match these terms with the correct statement or definition:

Ectopic pregnancy
Miscarriage
Placental abruption

Placenta previa
Pregnancy induced hypertension

_____ 1. Pregnancy that occurs outside the uterus.

_____ 2. Delivery of the fetus prior to 37 weeks of the pregnancy; before 24 weeks, usually results in the death of the fetus.

_____ 3. Implantation of the blastocyst near the opening into the cervical canal.

_____ 4. Tearing away of the placenta previa from the uterine wall.

_____ 5. Condition that results in sudden weight gain associated with edema; toxemia of pregnancy.

B. Match these terms with the
correct statement or definition:

Fetal alcohol effect   Teratogens
Fetal alcohol syndrome

_____   1. Drugs that can cross the placenta and cause birth defects in the
developing embryo.

_____   2. Child's condition caused by consumption of substantial amounts of
alcohol by a woman during pregnancy; includes brain dysfunction and
facial peculiarities.

_____   3. Child's condition caused by consumption of substantial amounts of
alcohol by a woman during pregnancy; includes brain dysfunction
without facial peculiarities.

Chromosomes

" *DNA molecules become visible as chromosomes during cell division.* **"**

A. Match these terms with the
correct statement or definition:

Deoxyribonucleic acid (DNA)   Somatic cells
Gametes

_____   1. The hereditary material of cells; also responsible for controlling cell
activities.

_____   2. All body cells except for sex cells, e.g., muscle cells or neurons.

_____   3. Sex cells, e.g., sperm cells or oocytes.

☞ Genetics is the study of heredity, that is, those characteristics inherited by children from
their parents.

B. Match these terms with the
correct statement or definition:

23   XX
23 pairs   XY

_____   1. Total number of chromosomes in a human somatic cell.

_____   2. Total number of chromosomes in gametes.

_____   3. Sex chromosomes in a normal female somatic cell.

_____   4. Sex chromosomes in a normal male somatic cell.

C. Match these terms with the
correct statement or definition:

Autosomal chromosomes   Meiosis
Karyotype   Sex chromosomes

_____   1. Display of chromosomes in a somatic cell.

_____   2. All chromosomes except X or Y; 22 pairs in a normal somatic cell.

_____   3. X or Y chromosomes; one pair in a normal somatic cell.

_____   4. Process by which gametes are produced from somatic cells.

D. Using the terms provided, complete the following statements:

Fertilization   X
Gamete   Y
Reduction

Meiosis is called _(1)_ division because the number of chromosomes in the gametes is half the number in somatic cells. When a sperm cell and an oocyte fuse during _(2)_, each contributes one half of the chromosomes necessary to produce new somatic cells. During meiosis, the chromosomes are distributed in such a way that each _(3)_ receives only one chromosome from each pair of chromosomes. For example, during gamete formation, the pair of sex chromosomes separates and an oocyte has only a(n) _(4)_ chromosome whereas a sperm cell has only an X chromosome or only a Y chromosome. If the oocyte is fertilized by a sperm with a(n) _(5)_ chromosome, a male results, but if the oocyte is fertilized by a sperm with a(n) _(6)_ chromosome, a female results.

1. _____
2. _____
3. _____
4. _____
5. _____
6. _____

Genes

“_The functional unit of heredity is the gene._**”**

A. Match these terms with the correct statement or definition:

Carrier   Homozygous
Dominant   Phenotype
Genotype   Recessive
Heterozygous

_____ 1. Two genes for a trait are identical.

_____ 2. Gene that masks the effect of another gene for the same trait.

_____ 3. The actual genes that a person possesses.

_____ 4. The expression of the genes that a person possesses.

_____ 5. A heterozygous person with an abnormal recessive gene.

B. Match these terms with the correct statement or definition:

AB blood   Hemophilia
Albinism   Sickle cell anemia
Height and skin color   Sickle cell trait

_____ 1. Caused by homozygous genes defective for melanin production

_____ 2. Caused by sex-linked recessive gene.

_____ 3. Caused by homozygous recessive genes for abnormal hemoglobin.

_____ 4. Caused by half normal and half abnormal hemoglobin.

_____ 5. Caused by two genes neither dominant nor recessive to each other.

_____ 6. Caused by multiple genes on different chromosomes.

Genetic Disorders

" *Approximately 15% of all congenital disorders have a known genetic cause.* "

A. Match these terms with the
correct statement or definition:

Congenital disorders
Genetic disorders

_____   1. Caused by abnormalities in DNA.

_____   2. Defects present at birth, regardless of cause.

B. Match these terms with the
correct statement or definition:

Chromosomal disorder
Mutation

_____   1. Random change in the number or kinds of nucleotides composing DNA.

_____   2. Results from abnormal distribution of chromosomes or parts of
chromosomes during gamete formation; e.g., Down syndrome.

C. Match these terms with the
correct statement or definition:

Cancer   Genetic predisposition
Carcinogens   Oncogenes

_____   1. Tumor resulting from uncontrolled cell division.

_____   2. Genes associated with cancer.

_____   3. Chemicals that initiate the development of cancer.

_____   4. Increased likelihood a person will develop a disorder; e.g., cancer.

Genetic Counseling and the Human Genome

" *Genetic counseling includes predicting the results of matings involving carriers of harmful genes.* "

Using the terms provided, complete the following statements:

Amniocentesis   Genotype
Chorionic villi sampling   Human genome
Gene therapy   Karyotype
Genomic map   Pedigree

A first step in genetic counseling is to determine the _(1)_ of the
individuals involved. A family tree or _(2)_ provides historical
information about family members. A _(3)_ can be taken from
white blood cells, or the amount of a substance produced by a
carrier can be tested. Fetal cells can be tested by _(4)_, which
takes cells floating in the amniotic fluid, or _(5)_, which takes
cells from the fetal side of the placenta. The _(6)_ is all the
genes found in human beings; a _(7)_ is a description of genes and
their location on the chromosomes. Knowledge of the genes
involved may result in _(8)_ to repair or replace defective genes,
but may raise legal and ethical questions.

1. _____

2. _____

3. _____

4. _____

5. _____

6. _____

7. _____

8. _____

1. List the following events or structures in the order in which they occur during development: blastocyst, embryo, fertilization, fetus, implantation, zygote.

2. Name the three germ layers from which all adult tissues develop.

3. Differentiate between the processes that produce fraternal twins and identical twins.

4. List the conditions associated with the three stages of labor.

5. List three hormones that increase uterine contractions.

6. Distinguish between gametes and somatic cells.

7. Explain the difference between a person's genotype and phenotype.

Place the letter corresponding to the correct answer in the space provided.

_____ 1. The major development of organs takes place in the
a. organ period.
b. fetal period.
c. blastocyst period.
d. embryonic period.

_____ 2. Given the following structures:
1. blastocyst
2. embryonic disk
3. zygote

Choose the arrangement that lists the structures in the order in which they are formed during development.
a. 1,2,3
b. 1,3,2
c. 2,3,1
d. 3,1,2
e. 3,2,1

_____ 3. The placenta
a. develops from the chorion.
b. allows maternal blood to mix with embryonic blood.
c. is attached to the embryo with chorionic villi.
d. all of the above

_____ 4. The embryo develops from the
a. inner cell mass.
b. wall of the blastocyst.
c. amniotic cavity.
d. yolk sac.

_____ 5. The brain develops from
a. endoderm.
b. ectoderm.
c. mesoderm.

_____ 6. Given the following events:
1. blastocyst formation
2. implantation
3. organ formation

Choose the arrangement that lists the events in the order in which they occur during development.
a. 1,2,3
b. 1,3,2
c. 2,1,3
d. 3,1,2
e. 3,2,1

_____ 7. During the fetal period
a. most organ systems enlarge and mature.
b. little increase in length or weight occurs.
c. ectoderm, endoderm, and mesoderm are produced.
d. all of the above

_____ 8. Which of the following hormones does not stimulate uterine muscle contraction?
a. estrogen
b. prostaglandins
c. oxytocin
d. human chorionic gonadotropin (HCG)

_____ 9. During the third stage of labor,
a. the cervix becomes dilated to 10 cm.
b. expulsion of the placenta from the uterus occurs.
c. the baby exits the vagina.

_____ 10. A pregnancy that occurs outside the uterus is called a(n)
a. teratogen.
b. placenta previa.
c. placental abruption.
d. ectopic pregnancy.
e. toxemia.

_____ 11. A normal male would have _____ pairs of autosomes and _____ sex chromosomes in each somatic cell.
a. 23,XY
b. 23,XX
c. 22,XY
d. 22,XX

_____ 12. A genetic disorder caused by a sex-linked recessive gene is
a. albinism.
b. hemophilia.
c. sickle cell anemia.
d. AB blood.
e. Down syndrome

_____ 13. A gene is
a. the functional unit of heredity.
b. a certain portion of a DNA molecule.
c. a part of a chromosome.
d. all of the above

_____ 14. Which of the following genotypes would be heterozygous?
a. DD
b. Dd
c. dd
d. both a and c

_____ 15. A carrier is a person with a
a. homozygous genotype of an abnormal gene.
b. homozygous genotype without an abnormal gene.
c. heterozygous genotype with an abnormal recessive gene.
d. heterozygous genotype without an abnormal recessive gene.

☆ ——— FINAL CHALLENGES ——— ☆

Use a separate sheet of paper to complete this section.

1. Considering the factors that influence parturition, explain why would it be likely that multiple births occur earlier than the 38 weeks which is usually considered full-term?

2. Although the placenta is important for exchanging nutrients, gases, and waste products between the mother and fetus, there is no actual mixing of maternal blood with fetal blood. Why is it necessary to keep maternal and fetal blood separate? (HINT: only half of the genetic makeup of the fetus comes from the mother)

3. What would be the result at fertilization if sperm cells and oocytes were produced without meiosis? What would happen with each succeeding generation?

4. A friend tells you that a male can only inherit hemophilia from his mother. Another friend, however, points out that a male still inherits half of his chromosomes from his father, and half of his chromosomes from his mother. How would you reconcile these two statements?

5. The gene for sickle cell hemoglobin is recessive (s) to the gene for normal hemoglobin (S). A man with sickle cell trait (Ss) and a woman with normal hemoglobin (SS) have a child. What is the probability the child will have sickle cell anemia? What is the probability the child will have the sickle cell trait?

6. A woman with polydactyly (extra fingers and/or toes) has surgery to remove the extra fingers and toes. Polydactyly is a dominant trait. Her father also had polydactyly, but her mother had the normal number of fingers and toes. Assuming that her husband has normal fingers and toes, is it possible for them to have a child with polydactyly? Is it possible for them to have a normal child?

7. Albinism is caused by a recessive gene (a). If a heterozygous normal person (Aa) mates with a person with albinism (aa), what percentage of their children would be expected to have albinism?

ANSWERS TO CHAPTER 20

CONTENT LEARNING ACTIVITY

Development
A. 1. Fertilization; 2. Zygote; 3. Blastocyst; 4. Inner cell mass; 5. Identical twins; 6. Fraternal twins
B. 1. Yolk sac; 2. Amniotic cavity; 3. Embryonic disk; 4. Primitive streak; 5. Embryo; 6. Fetus
C. 1. Germ layers; 2. Endoderm; 3. Mesoderm; 4. Ectoderm
D. 1. Uterine tube; 2. Implantation; 3. Chorion; 4. Placenta; 5. Umbilical cord; 6. Chorionic villi; 7. Human chorionic gonadotropin (HCG)

Parturition
A. 1. First stage of labor; 2. Second stage of labor; 3. Third stage of labor
B. 1. Adrenocorticotropic hormone (ACTH); 2. Adrenal steroids; 3. Prostaglandins; 4. Oxytocin; 5. Contractions

Disorders of Pregnancy
A. 1. Ectopic pregnancy; 2. Miscarriage; 3. Placenta previa; 4. Placental abruption; 5. Pregnancy induced hypertension
B. 1. Teratogens; 2. Fetal alcohol syndrome; 3. Fetal alcohol effect

Chromosomes

A. 1. Deoxyribonucleic acid (DNA); 2. Somatic cells; 3. Gametes
B. 1. 23 pairs; 2. 23; 3. XX; 4. XY
C. 1. Karyotype; 2. Autosomal chromosomes; 3. Sex chromosomes; 4. Meiosis
D. 1. Reduction; 2. Fertilization; 3. Gamete; 4. X; 5. Y; 6. X

Genes
A. 1. Homozygous; 2. Dominant; 3. Genotype; 4. Phenotype; 5. Carrier
B. 1. Albinism; 2. Hemophilia; 3. Sickle cell anemia; 4. Sickle cell trait; 5. AB blood; 6. Height and skin color

Genetic Disorders
A. 1. Genetic disorders; 2. Congenital disorders
B. 1. Mutation; 2. Chromosomal disorder
C. 1. Cancer; 2. Oncogenes; 3. Carcinogens; 4. Genetic predisposition

Genetic Counseling and the Genome
1. Genotype; 2. Pedigree; 3. Karyotype; 4. Amniocentesis; 5. Chorionic villi sampling; 6. Human genome; 7. Genomic map; 8. Gene therapy

QUICK RECALL

1. Fertilization, zygote, blastocyst, implantation, embryo, fetus
2. Ectoderm, endoderm, and mesoderm.
3. Identical twins: fertilization of an oocyte by a sperm cell, followed by separation of the developing cells into two individuals; Fraternal twins: two oocytes fertilized by two different sperm cells.
4. First stage of labor: onset of regular contractions until cervix is dilated 10 cm; second stage of labor: maximal cervical dilation until baby exits vagina; third stage of labor: expulsion of the placenta.
5. Estrogen, prostaglandins, and oxytocin.
6. Gametes: sex cells with 23 chromosomes; somatic cells: all cells except sex cells with 23 pairs of chromosomes.
7. Genotype: the actual genes a person possesses; phenotype: the expression of the genes a person possesses.

MASTERY LEARNING ACTIVITY

1. D. Most organ development takes place in the embryonic period (second to eighth week). The germ layers develop from the blastocyst (first two weeks), and growth and maturation occur in the fetal period (the last seven months).

2. D. Fertilization of the oocyte produces a zygote, which divides and develops into the blastocyst. After implantation of the blastocyst, further division produces the amniotic cavity and forms the embryonic disk.

3. A   The placenta develops from the chorion. There is no mixing of maternal and fetal blood, and the placenta is attached to the embryo by the umbilical cord.

4. A   The embryo develops from the inner cell mass. The chorion develops from the wall of the blastocyst; the amniotic cavity and yolk sac are spaces associated with the embryo.

5. B   The brain, spinal cord, and nerves develop from ectoderm.

6. A   The blastocyst (3-4 days) is formed before implantation (7 days), followed by organ formation (2-8 weeks).

7. A   Most of the organ systems are formed during the embryonic period, and the fetal period is a time of growth and maturation of those organ systems.

8. D   Human chorionic gonadotropin (HCG) stimulates the corpus luteum to produce estrogen and progesterone. All of the other hormones listed stimulate uterine muscle contraction.

9. B   During the third stage of labor the placenta is expelled. During the first stage of labor, the cervix becomes dilated to 10 cm, and during the second stage of labor the baby exits the vagina.

10. D   A pregnancy that occurs outside the uterus is an ectopic pregnancy; this occurs most frequently in the uterine tube. A teratogen is a chemical that passes across the placenta and causes birth defects. A placenta previa is implantation of the blastocyst close to the cervix; if the placenta previa tears away from the uterine wall, it is a placental abruption. Toxemia is another name for pregnancy induced hypertension.

11. C   A normal male would have 22 pairs of autosomes and XY sex chromosomes in each somatic cell.

12. B   Hemophilia is caused by a sex-linked recessive gene. Albinism and sickle cell anemia are caused by recessive genes that are not sex-linked. AB blood is a product of two genes which are neither dominant or recessive to each other, and Down syndrome is a result of a chromosomal disorder (an extra #21 chromosome).

13. D   All of the statements correctly describe a gene.

14. B   Dd is heterozygous. DD or dd are homozygous genotypes.

15. C   A heterozygous genotype produces a normal phenotype, even though there is an abnormal gene present. Therefore, the person does not show the trait, even though they possess a gene for that trait.

 FINAL CHALLENGES

1. Multiple fetuses are crowded in the uterus much earlier than a single fetus. Stress from crowding increases the production of ACTH in the pituitary of the fetus. ACTH stimulates release of adrenal steroids that cause the placenta to increase production of estrogen and prostaglandins, both of which increase uterine contractility. As a result of earlier stress from crowding, parturition most likely occurs earlier, also.

2. Lymphocytes in the mother's blood recognize and respond to foreign antigens. Rejection is mostly a result of T cells; in this case, fetal tissues, as they are formed, are constantly exposed to T cells in the mother's blood. This results in a rejection process that destroys the fetal tissues. Also, if the mother and the fetus are different blood types, a transfusion reaction occurs between them . Therefore, a direct connection between the two circulatory systems does not allow normal development of the fetus to occur.

3. If sperm cells and oocytes were produced without meiosis, they would each have 23 pairs of chromosomes, as in somatic cells. When the sperm cell and oocyte combined during fertilization, the zygote would then have 46 pairs of chromosomes, or double the normal number. With each succeeding generation, the number of chromosomes would be doubled. In humans, extra chromosomes almost always cause defects; in addition, in a few generations the whole cell would be filled with chromosomes.

4. A male does inherit hemophilia only from his mother. To be a male, he has to inherit a Y chromosome, (which can only come from his father), and an X chromosome (which therefore must come from his mother). Because the recessive gene for hemophilia is only carried on the X chromosome, a male always inherits the gene from his mother, even though he does get one sex chromosome from each parent.

5. Remember that the genotype "ss" produces sickle cell anemia, and the genotype "Ss" produces the sickle cell trait. The woman (genotype = SS) only contributes a normal gene to her children. The man is heterozygous (Ss); half of his gametes have a normal hemoglobin gene, and half have the sickle cell gene. If a sperm cell with a normal gene combines with an oocyte with the normal gene, the children would be normal. If a sperm with the sickle cell gene combines with an oocyte with normal gene, the child would have sickle cell trait. Therefore, the probability is 50% that the child will have sickle cell trait. There is no probability that the child will have sickle cell anemia because the woman always contributes a normal gene to her children.

6. Surgery only removes the extra fingers and toes, but does not affect the woman's genotype. She must be heterozygous for polydactyly (Pp), because her mother was normal. Therefore, half of the oocytes she produces will have the gene for polydactyly. Her husband has normal fingers and toes, and therefore is homozygous normal (genotype = pp). The prediction is that half of their children would have polydactyly, and half would be normal.

7. One parent is heterozygous (Aa), and the parent with albinism is homozygous (aa). Based on probability, half of their children would have albinism (aa), and the other half would be carriers (Aa).

Infectious Diseases

FOCUS: Microorganisms include bacteria, viruses, fungi, and protozoans. Some microorganisms are responsible for the development of diseases in humans. Understanding the structures and processes of microorganisms is beneficial because such knowledge allows us to prevent and treat diseases. For example, one can attack parts of bacteria that are not found in humans or one can interfere with the reproductive process of viruses. Microorganisms are found in nonliving reservoirs, such as water, food, and soil, and in living reservoirs, such as humans and animals. Medical techniques and public health measures are designed to prevent the transmission of disease-causing microorganisms from reservoirs to noninfected humans.

CONTENT LEARNING ACTIVITY

Disease Terminology

66*Disease is defined as an interference with normal body functions such that*99 *homeostasis is not maintained.*

A. Match these terms with the correct statement or definition:

Diagnosis Sign
Host Symptom
Parasite Therapy
Prognosis

_____ 1. Organism that lives on or in another organism.

_____ 2. Any departure from normal that the health care provider can objectively observe.

_____ 3. Statement by a patient, "I feel hot."

_____ 4. Body temperature is 3 degrees above normal.

_____ 5. Identifying the disease.

_____ 6. Estimation of the outcome of a disease.

B. Match these terms with the
correct statement or definition:

Acute   Infection
Chronic   Local
Endemic   Prevalence
Epidemic   Subacute
Incidence   Systemic
Incubation period

_____ 1. Invasion and multiplication of a pathogen within the body.

_____ 2. Time between the entry of a pathogen into the body and the first appearance of symptoms.

_____ 3. Disease that lasts a short time and usually has severe symptoms.

_____ 4. Infection that spreads throughout the body.

_____ 5. Number of new cases of a disease that appear in a given period.

_____ 6. Number of old and new cases at a given time.

_____ 7. Disease that is continuously present in a population.

Bacteria

"Bacteria are single-celled organisms that occur almost everywhere."

A. Match these terms with the
correct statement or definition:

Aerobic   Coccus
Anaerobic   Staphylococcus
Bacillus   Streptococcus

_____ 1. Bacteria that can survive in the absence of oxygen.

_____ 2. Rod-shaped bacteria.

_____ 3. Long chain of spherical-shaped bacteria.

_____ 4. Cluster of spherical-shaped bacteria.

☞ Bacteria exist in one of three basic shapes: bacillus, coccus, or spiral shape.

B. **M**atch these terms with the correct statement or definition:

Capsule
Cell wall
Flagellum
Lipopolysaccharide layer

Lysozyme
Peptidoglycan
Pilus

_____ 1. Gelatinous, sticky material that surrounds a bacteria; helps them stick to surfaces.

_____ 2. Short, hollow tube that functions to attach bacteria to surfaces or transfer genetic material to other bacteria.

_____ 3. Long, corkscrew-shaped structure that rotates to move bacteria.

_____ 4. Nonliving, rigid structure responsible for the shape of bacteria.

_____ 5. Molecules found in bacterial cell walls; provides structural strength.

_____ 6. Enzyme in saliva, tears, and phagocytes that destroys peptidoglycan.

_____ 7. Gram negative bacteria have this substance in their cell walls, whereas gram positive bacteria do not.

_____ 8. Pieces of this substance become endotoxins, which produce the symptoms of fever, aches, weakness, and sometimes shock.

C. **M**atch these terms with the correct statement or definition:

Endospore
Exotoxin
Nuclear area

Plasma membrane
Plasmid
Ribosome

_____ 1. Outer, living boundary of a bacterial cell.

_____ 2. Part of a bacterial cell containing a single, long circular molecule of DNA; controls cell activities.

_____ 3. Short, circular molecules of DNA not normally required for cell growth and survival.

_____ 4. Bacterial structure that is a site of protein synthesis; structurally different from those found in humans.

_____ 5. Proteins released from bacteria that are responsible for disease symptoms.

_____ 6. Dormant form of bacteria that is resistant to destruction.

D. Match these terms with the correct parts labeled in Figure : 21-1.

Capsule
Cell wall
Flagella
Nuclear area

Pili
Plasma membrane
Plasmid
Ribosome

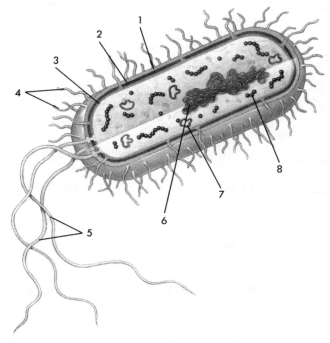

1. _____   4. _____   7. _____

2. _____   5. _____   8. _____

3. _____   6. _____

E. Match these terms with the correct statement or definition:

Antibody tests
Asexual reproduction
Growth

Metabolic tests
Staining tests

_____   1. Bacterial process that produces two genetically identical daughter cells.

_____   2. Increase in the number of bacteria.

_____   3. First step in bacterial identification considers this characteristic in addition to bacterial shape and organization.

_____   4. Identify bacteria by determining their nutritional requirements.

_____   5. Identify bacteria using specific proteins that bind to only one type of bacteria.

320

Understanding Genetic Transfer and Recombination

❝_Bacteria can acquire new characteristic by receiving DNA they do not already have._**❞**

Using the terms provided, complete the following statements:

Genetic engineering	Sex pilus
Plasmid	Sexual reproduction
Recombination	

Transfer of DNA between bacteria can occur through a hollow tube called a _(1)_. Part of a chromosome or a _(2)_ moves from one bacterial cell to another. This process is sometimes called _(3)_. The insertion of the DNA from the chromosome or plasmid into the chromosome of the recipient cell is called _(4)_. The transferred DNA can function to give the bacterial cell new characteristics. Recombinant DNA technology, or _(5)_, is the deliberate combining of DNA by humans.

1. _____

2. _____

3. _____

4. _____

5. _____

Viruses

❝_All viruses intracellular, existing as some time inside a living cell._**❞**

A. Match these terms with the correct statement or definition:

Capsid	Inert particle
DNA or RNA	Reproduce
Envelope	Spikes

_____ 1. What a virus is outside of a living cell.

_____ 2. What a virus does inside a living cell.

_____ 3. Types of nucleic acids inside viruses; responsible for viral reproduction.

_____ 4. Protein coat surrounding the nucleic acids of viruses; protects the nucleic acids and attaches the virus to cells.

_____ 5. Lipoprotein structure that surrounds the capsid of some viruses; protects the nucleic acids and attaches the virus to cells.

_____ 6. Projections from the envelope.

B. Match these terms with
the correct parts labeled
in Figure : 21-2

Capsid
Envelope
Nucleic acid
Spikes

1. _____

2. _____

3. _____

4. _____

C. Match these terms with the
correct statement or definition:

Attachment   Release
Entry   Uncoating
Production

_____ 1. Envelope or capsid binds to a cell membrane.

_____ 2. Cell takes in the virus by endocytosis.

_____ 3. Envelope and capsid are broken down.

_____ 4. Viral nucleic acid directs the cell to produce viral nucleic acids and
proteins.

_____ 5. Newly formed viruses pass through the cell membrane.

☞ To kill viruses it may be necessary to kill the cells they infect. Antiviral drugs prevent nucleic
acids from functioning.

Additional Organisms

❝Organisms other than bacteria and viruses cause disease.❞

Match these terms with the
correct statement or definition:

Fungi
Parasitic worms
Protozoans

_____ 1. Single-celled organisms that are structurally like animal cells.

_____ 2. Plantlike single-celled or multicellular organisms.

_____ 3. Multicellular animals that spend part of their lives in humans.

322

The Spread of Pathogens

"Infection and disease can result when pathogens are introduced into the body."

A. Match these terms with the correct statement or definition:

Carrier   Nonliving reservoir
Living reservoir   Reservoir

_____ 1. General term for the source of a disease-producing organism.

_____ 2. Water, food, and soil are examples.

_____ 3. Humans and animals are examples.

_____ 4. Infected person who releases pathogens before the symptoms of disease appear or for a short time following recovery from the disease.

B. Match these terms with the correct statement or definition:

Direct contact   Portal
Indirect contact   Vector
Opportunistic pathogen

_____ 1. Transmission in which there is a close association between the reservoir and the human infected.

_____ 2. Transmission by contaminated objects, contaminated food or water, and vectors.

_____ 3. Examples include sexual intercourse, kissing, handshaking, placental transmission, and animal bites.

_____ 4. Examples include dust particles or dried nasal secretions,

_____ 5. Invertebrate (without a backbone) animal that transmits a disease.

_____ 6. Point where a pathogen enters or exits the body.

_____ 7. Microorganism that takes up residence in the body but does not cause disease until conditions become favorable.

Preventing Disease

" *Diseases do not occur if pathogens are not allowed to infect the body.* **"**

Match these terms with the
correct statement or definition:

Antisepsis　　　　　　　Sterilization
Disinfection　　　　　　Surgical asepsis
Medical asepsis

_____ 1. Complete destruction of all organisms.

_____ 2. Destruction of pathogens on inanimate objects.

_____ 3. Destruction of pathogens on the skin or tissues.

_____ 4. Autoclaving, baking, burning, filtration, radiation, and treatment
with ethylene oxide are examples.

_____ 5. Keeps all microorganisms away from an object or person.

_____ 6. Keeps away pathogens associated with communicable diseases;
quarantine is an example.

☞ Phagocytes and white blood cells of the immune system destroy pathogens after they enter the
body. Chemotherapy uses drugs to kill or suppress pathogens that have entered the body.

Public Health

" *Public health measures are the best method of preventing the outbreak of epidemics.* **"**

Match these terms with the
correct statement or definition:

Animal control　　　　　Sewage disposal
Food safety　　　　　　　Water safety

_____ 1. Cooking pork to prevent trichinosis is an example.

_____ 2. Filtration and disinfection with chlorine is an example.

_____ 3. Separating sewage into effluent and sludge.

_____ 4. Vaccinating dogs against rabies.

_____ 5. Decreasing the size of animal populations that are reservoirs for
human diseases.

Nosocomial Infections

66 *Although medical therapy is intended to cure, sometimes it causes disease.* **99**

Using the terms provided, complete the following statements:

Carriers   Nosocomial

Contaminated   Weakened

Direct contact

Infections that the patient acquires as a result of being in a hospital or other health-related facility are called _(1)_ infections. These infections can occur because the patient is in a _(2)_ condition, because they are exposed to health personnel or other patients who are _(3)_, or because they are exposed to _(4)_ objects. In addition, medical procedures can transmit pathogens through _(5)_.

1. _____
2. _____
3. _____
4. _____
5. _____

QUICK RECALL

1. List the shapes and organizational relationships of bacteria.

2. Contrast endotoxins and exotoxins.

3. List three types of tests used to identify bacteria.

4. Contrast asexual and sexual reproduction in bacteria.

5. List the five steps of viral replication.

6. List the two general categories of reservoirs and the two general types of transmission.

7. List the four portals.

8. List the four major public health measures.

MASTERY LEARNING ACTIVITY

Place the letter corresponding to the correct answer in the space provided.

_____1. Any departure from normal objectively determined by a doctor.
   a. diagnosis
   b. prognosis
   c. sign
   d. symptom

_____2. An infection that lasts a short time and affects one area of the body.
   a. acute and local
   b. acute and systemic
   c. chronic and local
   d. chronic and systemic

_____3. Staphylococci are
   a. clusters of spherical-shaped bacteria.
   b. long chains of spherical-shaped bacteria.
   c. individual spherical-shaped cells with no relationship to each other.
   d. none of the above

_____4. Which of the following structures enable bacteria to stick to surfaces?
   a. capsule
   b. pilus
   c. flagellum
   d. a and b
   e. all of the above

_____5. Bacteria that have this chemical stain red and are said to be gram negative.
   a. exotoxin
   b. lipopolysaccharide layer
   c. lysozyme
   d. peptidoglycan

_____6. Which of the following structures is found in bacteria?
   a. endoplasmic reticulum
   b. mitochondria
   c. nucleus
   d. none of the above

_____7. Plasmids
   a. are circular molecules of DNA.
   b. form the nuclear area of bacteria.
   c. are a site of protein synthesis in
      bacteria.
   d. regulate the movement of materials
      into and out of bacteria.

_____8. Bacterial growth
   a. means individual bacterial cells
      become larger.
   b. refers to asexual reproduction in
      bacteria.
   c. involves the transfer of DNA
      between bacteria using a pilus.
   d. requires plasmids.

_____9. It has been determined that bacteria A
      requires glucose in order to replicate.
      Along with other information, this fact
      is used to identify the bacteria and is an
      example of
   a. antibody testing.
   b. gram stain testing.
   c. metabolic testing.
   d. reproductive testing.

_____10. Lipoproteins responsible for the
      attachment of viruses to cells.
   a. capsule
   b. envelope
   c. peptidoglycan
   d. spore coat

_____11. Stage of viral reproduction in which
      viral proteins form around viral nucleic
      acids.
   a. attachment
   b. entry
   c. production
   d. release
   e. uncoating

_____12. Which of the following statements is
      true?
   a. Direct reservoirs are other human
      beings.
   b. Indirect reservoirs are objects, food,
      or water.
   c. A vector is an invertebrate animal
      that transmits a disease.
   d. all of the above

_____13. Concerning portals,
   a. they can be entered through direct
      contact or indirect contact.
   b. there are four of them.
   c. the respiratory tract,
      gastrointestinal tract, urogenital
      tract, and skin are examples.
   d. the portal of entry is often, but not
      always, the portal of exit.
   e. all of the above

_____14. Opportunistic pathogens can cause
      disease
   a. when transferred from one part of
      the body to another part.
   b. when the host is weakened by
      another disease.
   c. following destruction of normal,
      nonpathogenic microorganisms of the
      body.
   d. all of the above

_____15. Antisepsis
   a. completely destroys all
      microorganisms.
   b. is the destruction of pathogens on
      inanimate objects.
   c. is the destruction of pathogens on the
      skin.
   d. keeps away pathogens associated
      with communicable diseases.

_____16. Proper disposal of garbage is an
      example of
   a. food safety.
   b. water safety.
   c. sewage disposal.
   d. animal control.

FINAL CHALLENGES

Use a separate sheet of paper to complete this section.

1. In what ways are virus like nonliving matter and like living organisms?

2. Viruses often infect specific cell types, but not other cells. For example, the AIDS virus infects certain white blood cells. Propose an explanation (Hint: hormones).

3. A doctor at a hospital often givens antibiotic injections to treat certain infections. Prior to giving the shot, he clears the syringe of air. As the air is removed, some of the antibiotic is ejected into the air. At this same hospital, a number of surgical patients developed infections with bacteria that were resistant to the antibiotic used by the doctor. Explain how this could happen.

4. A woman who frequently used a commercially prepared douche developed a bacterial infection of the vagina. Propose some ways in which this could have happened.

5. Everyone who attended a company picnic develop nausea, vomiting, and diarrhea. Coworkers who did not attend the picnic did not become ill. Propose an explanation and describe the possible reservoir, mode of transmission, and portal of entry involved.

6. A small rural town has a water treatment and sewage disposal facility. Two weeks following a flood, there was an outbreak of diarrhea in the town. Explain what happened. Why did it take two weeks?

ANSWERS TO CHAPTER 21

CONTENT LEARNING ACTIVITY

Disease Terminology
A. 1. Parasite; 2. Sign; 3. Symptom; 4. Sign; 5. Diagnosis; 6. Prognosis
B. 1. Infection; 2. Incubation period; 3. Acute; 4. Systemic; 5. Incidence; 6. Prevalence; 7. Endemic

Bacteria
A. 1. Anaerobic; 2. Bacillus; 3. Streptococcus; 4. Staphylococcus
B. 1. Capsule; 2. Pilus; 3. Flagellum; 4. Cell wall; 5. Peptidoglycan; 6. Lysozyme; 7. Lipopolysaccharide layer; 8. Lipopolysaccharide layer
C. 1. Plasma membrane; 2. Nuclear area; 3. Plasmid; 4. Ribosome; 5. Exotoxin; 6. Endospore
D. 1. Capsule; 2. Cell wall; 3. Plasma membrane; 4. Pili; 5. Flagella; 6. Nuclear area; 7. Plasmid; 8. Ribosome
E. 1. Asexual reproduction; 2. Growth; 3. Staining tests; 4. Metabolic tests; 5. Antibody tests

Understanding Genetic Transfer and Recombination
1. Sex pilus; 2. Plasmid; 3. Sexual reproduction; 4. Recombination; 5. Genetic engineering

Viruses
A. 1. Inert particle; 2. Reproduce; 3. DNA or RNA; 4. Capsid; 5. Envelope; 6. Spikes
B. 1. Spikes; 2. Envelope; 3. Capsid; 4. Nucleic acid
C. 1. Attachment; 2. Entry; 3. Uncoating; 4. Production; 5. Release

Additional Organisms
1. Protozoans; 2. Fungi; 3. Parasitic worms

The Spread of Pathogens
A. 1. Reservoir; 2. Nonliving reservoir; 3. Living reservoir; 4. Carrier
B. 1. Direct contact; 2. Indirect contact; 3. Direct contact; 4. Indirect contact; 5. Vector; 6. Portal; 7. Opportunistic pathogen

Preventing Disease
1. Sterilization; 2. Disinfection; 3. Antisepsis; 4. Sterilization; 5. Surgical asepsis; 6. Medical asepsis

Public Health

1. Food safety; 2. Water safety; 3. Sewage disposal;
4. Animal control; 5. Animal control

Nosocomial Infections

1. Nosocomial infections; 2. Weakened; 3. Carriers;
4. Contaminated; 5. Direct contact

QUICK RECALL

1. Shapes: bacillus (rod-shaped), coccus (spherical-shaped), and spiral shaped. Organizational: streptococcus (chain of cocci) and staphylococcus (cluster of cocci)

2. Endotoxins are lipopolysaccharides derived from the LPS layer of gram negative bacteria after the bacteria die. Exotoxins are proteins produced and released from living bacteria (gram negative or gram positive)

3. Staining test (also determines shape and organization), metabolic test, and antibody tests

4. Asexual reproduction is the division by mitosis of one bacterial cell to form two genetically identical daughter cells. Sexual reproduction is the transfer of DNA between two bacterial cells through a pilus; the process begins and ends with two cells that are not genetically identical

5. Attachment, entry, uncoating, production, and release

6. Reservoirs: living and nonliving; transmission: direct and indirect

7. Respiratory tract, digestive tract, urogenital tract, and skin

8. Food safety, water safety, sewage disposal, and animal control

MASTERY LEARNING ACTIVITY

1. C. A sign is any departure from normal objectively determined, whereas a symptom is a departure subjectively experienced by the patient. A diagnosis is the determination of the cause of the disease, whereas a prognosis is the projected outcome of the disease.

2. A. Acute infections last a short time and usually have severe symptoms, whereas chronic infections last a long time and can progressively become severe. Local infections occur in one part of the body, whereas systemic infections spread throughout the body.

3. A. Staphylococci are clusters of cocci (spherical-shaped bacteria), whereas streptococci are chains of cocci.

4. D. The capsule and pilus enable bacteria to stick to surfaces. The flagellum enables bacteria to move about.

5. B. The lipopolysaccharide layer, which results in a red stain, is a characteristic of gram negative bacteria. Gram positive bacteria, which do not have the lipopolysaccharide layer, stain purple.

6. D. None of the structures are found in bacteria, although they are all found in human cells.

7. A. Plasmids are small, circular molecules of DNA. The chromosome, which is a large circular molecule of DNA, forms the nuclear area. Ribosomes are the site of protein synthesis, and the plasma membrane regulates the movement of materials into and out of bacteria.

8. B. Bacterial growth refers to an increase in the number of bacteria as a result of cell divisions by mitosis (asexual reproduction). The transfer of DNA between bacteria using a pilus is sometimes called sexual reproduction.

9. C. Metabolic testing examines the nutritional requirements of bacteria.

10. B. The envelope is lipoproteins that can attach viruses to surfaces. The capsule attaches bacteria to surfaces. Peptidoglycan makes the cell wall of bacteria structurally stronger, and the spore coat protects the chromosome of an endospore from harsh environmental conditions.

11. C. Viral nucleic acids are surrounded by proteins (the capsid) to form new viruses during the production stage of viral replication.

12. C. A vector is an invertebrate (without a backbone) animal that transmits disease to humans. Reservoirs are living or not living, and transmission is by direct or indirect means.

13. D.   All of the statements are true.

14. D.   All of the statements are true.

15. C.   Antisepsis is the destruction of pathogens on the skin. Sterilization is the complete destruction of all microorganisms, disinfection is the destruction of pathogens on inanimate objects, and medical asepsis keeps away pathogens associated with communicable diseases.

16. D.   Animal control because animals feed on garbage, and reducing the amount of garbage reduces the size of animal populations that can spread disease to humans.

 FINAL CHALLENGES

1. Nonliving: outside of living cells viruses are inert particles. Living organisms: viruses are capable of reproduction and directing cellular activities. Their nucleic acids can replicate and direct the synthesis of proteins, both of which are necessary for the formation of new viruses.

2. Hormones affect specific target tissues because the hormone binds to receptors on the target tissue. Tissues without the receptor are not affected by the hormone. The hormone and the target tissue receptor are specific for each other because of the manner in which they bind together. This is like the lock-and-key model of enzyme action. Viruses infect specific cells because they can attach only to cells that have specific receptors.

3. The ejected antibiotic goes into the air and is inhaled by health personnel, such as a nurse. The continual exposure to the antibiotic results in the development of antibiotic resistant bacteria in the respiratory tract of the nurse. She becomes a carrier for the resistant bacteria and inadvertently transmits the bacteria to the surgical patients.

4. There are several possibilities: 1) the bacteria were in the douche fluid; 2) performing the douche transmitted the bacteria from her hands, external genitalia, or perineum into the vagina; 3) the douche fluid changed the environment of the vagina or damaged the lining of the vagina, allowing the bacteria (opportunistic pathogen) to flourish.

5. It is reasonable to believe that something happened at the picnic, because coworkers who did not go to the picnic did not become ill. All of the symptoms are associated with the gastrointestinal tract. It is possible that contaminated food (nonliving reservoir), such as potato salad, was eaten (indirect contact) and entered the gastrointestinal portal.

6. The flooding allowed some of the water from the sewage disposal plant to get into the drinking water in the water treatment plant. People drinking the water were infected with a microorganism that causes diarrhea. There was a two week incubation period before symptoms appeared.